Archibald Geikie

Elementary Lessons in Physical Geography

Archibald Geikie

Elementary Lessons in Physical Geography

ISBN/EAN: 9783337277130

Printed in Europe, USA, Canada, Australia, Japan

Cover: Foto ©berggeist007 / pixelio.de

More available books at **www.hansebooks.com**

ELEMENTARY LESSONS

IN

PHYSICAL GEOGRAPHY

BY

ARCHIBALD GEIKIE, LL.D., F.R.S.

Director-General of the Geological Survey of the United Kingdom, and Director of the Museum of Practical Geology, Jermyn Street, London; formerly Murchison Professor of Geology and Mineralogy in the University of Edinburgh

ILLUSTRATED WITH WOODCUTS AND TEN PLATES

London
MACMILLAN AND CO.
AND NEW YORK
1892

PREFACE.

A SIMPLE but systematic and yet interesting description of the familiar features of the Earth's surface may be regarded as a natural introduction to the teaching of science. Treated from this point of view, Physical Geography may be made a valuable instrument of education. To give it such importance, the most advantageous method is to make use of the common knowledge and experience of the pupils, and, starting from this groundwork, to train them in habits of observation and in scientific modes of thought and inquiry among every-day phenomena. From the very outset the instruction should be as far as possible practical. A shower of rain, the growth and disappearance of a cloud, the flow of a brook, the muddy water of a river, the shape of a cliff, the outlines of a mountain, the undulations of a plain—these and the thousand other common features of landscape should be eagerly seized by the teacher and used as vivid illustrations of the broad fundamental principles which it will be his main object that his pupils should thoroughly master. Thus employed, Physical Geography is not learnt as an ordinary task, but rather becomes a delightful recreation, in which, however, the observing faculty is exercised, the power of induction cultivated, and the imagination kept constantly active.

Having been long convinced that such a method of instruction would place this branch of science upon a firmer and broader footing in our educational system, and would moreover prove of great service in fostering a spirit of observation and reflection even among children, I projected many years ago the *Primer of Physical Geography*, published in the series of *Science Primers*. The continued sale of large impressions of that little work encourages the hope that the method advocated has been found successful in practice.

The present volume may be regarded as a further development of the same plan of instruction. As its title implies, it still deals mainly with the broad elementary questions of Physical Geography. It would have been impossible to find a place within its Lessons for the treatment of every branch of the wide subject, and as impossible, had it even been desirable, to bestow equal fulness upon every branch for which room has been made. I have devoted most space to those aspects of the science which, in my own experience, have been found best suited for practical instruction. While as much general information as may be feasible should be communicated to them, young pupils cannot of course be expected to find the same interest in all divisions of the subject. It is of far more consequence to awaken in them a taste for such pursuits, and lead them to carry on the study of their own accord, than to try to charge their memories with dry facts and figures which, in the absence of intelligent and suggestive teaching, are too commonly meaningless and repulsive.

While, therefore, adhering to a systematic treatment, I have been led to dwell, for example, on the phenomena of the atmosphere at much greater length than is usual in elementary class-books. These phenomena are among the most familiar and universal features of the globe; examples of them can be constantly adduced, and they may thus be used with singular advantage to illustrate how the facts of science are observed and its laws are deduced.

I acknowledge with pleasure my obligations to my friend Mr. Buchan, who not only kindly allowed me to make use of his Charts of Atmospheric Pressure and Temperature, but who also read over the proof-sheets of the first two chapters and gave me valuable suggestions on subjects regarding which he is so high an authority.

The present Edition has been carefully revised throughout. Two new Plates have been added, one (from the "Challenger" results) showing the form of the great sea-basins as revealed by recent deep-sea research; the other (from Professor Loomis' Chart) indicating the general distribution of rain over the globe.

May 1884.

CONTENTS.

	PAGE
INTRODUCTION	1

CHAPTER I.—THE EARTH AS A PLANET.

LESSON	I. The Earth's Form	8
,,	II. The Earth's Motions	11
,,	III. The Earth and the Sun	17
,,	IV. Measurement and Mapping of the Earth's Surface	24
,,	V. A General View of the Earth	32

CHAPTER II.—AIR.

LESSON	VI. Its Composition	38
,,	VII. The Height of the Air	45
,,	VIII. The Pressure of the Air	47
,,	IX. The Temperature of the Air	54
,,	X. The Moisture of the Air	64
,,	XI. The Movements of the Air	83

CHAPTER III.—THE SEA.

LESSON	XII. The Great Sea-basins	103
,,	XIII. The Saltness of the Sea	113
,,	XIV. The Depths of the Sea	119
,,	XV. The Temperature of the Sea	125
,,	XVI. The Ice of the Sea	129
,,	XVII. The Movements of the Sea	137
,,	XVIII. The Offices of the Sea	152

Chapter IV.—THE LAND.

			PAGE
Lesson	XIX.	Continents and Islands	162
,,	XX.	The Relief of the Land—Mountains, Plains, and Valleys	173
,,	XXI.	The Composition of the Earth	182
,,	XXII.	Volcanoes	196
,,	XXIII.	Movements of the Land	210
,,	XXIV.	The Waters of the Land—Part I.—Springs and underground Rivers	222
,,	XXV.	The Waters of the Land—Part II.—Running Water—Brooks and Rivers	244
,,	XXVI.	The Waters of the Land—Part III.—Lakes and Inland Seas	258
,,	XXVII.	The Waters of the Land—Part IV.—The Work of Running Water	272
,,	XXVIII.	The Waters of the Land—Part V.—Frost, Snow-fields, Glaciers	293
,,	XXIX.	The Sculpture of the Land	314

Chapter V.—LIFE.

Lesson	XXX.	The Geographical Distribution of Plants and Animals	328
,,	XXXI.	The Diffusion of Plants and Animals—Climate—Migration and Transport—Changes of Land and Sea	337

Index 357

LIST OF ILLUSTRATIONS.

PLATES.

		TO FACE PAGE
Plate	I. The World in Hemispheres	24
,,	II. Pressure of the Atmosphere over the Globe in January	46
,,	III. Pressure of the Atmosphere over the Globe in July	50
,,	IV. Isothermal Lines over the Globe for January	54
,,	V. Isothermal Lines over the Globe for July	60
,,	VI. Map showing the mean Annual Rainfall of the Globe	76
,,	VII. The Ocean Basins	103
,,	VIII. Ocean Currents	144
,,	IX. Distribution of Earthquakes and Volcanoes over the Globe	200
,,	X. Zoological Provinces of the Globe	328

WOOD-CUTS.

FIG.		PAGE
	Ocean-Waves . . . . *Frontispiece*	
1.	Curvature of the Earth's Surface, as shown by ships at sea	9
2.	The Earth and Moon as seen from space	10
3.	The Earth's path round the Sun	14
4.	Plan of Volcanic Hills and Craters in the Bay of Naples	19
5.	A part of the Surface of the Moon, showing Volcanic Craters	20
6.	Measurement and Mapping of a Country by means of Triangulation	31
7.	What is seen after some raindrops collected in a town are evaporated, and the residue is placed below a microscope	40

LIST OF ILLUSTRATIONS.

FIG.		PAGE
8.	Diagram showing the influence of the varying thickness of the atmosphere in retarding the Sun's heat	57
9.	Clouds condensed by the Cliff of the Noss Head, Shetland, with a clear sky and a S.E. wind	73
10.	Snow-flakes	80
11.	Position and height of the Snow-line between Equatorial Africa and the North Polar Seas	81
12.	Sand-dunes—ridges of dry sand blown inland off the shore by the wind	100
13.	What is seen when a drop of concentrated sea-water is evaporated under a microscope	115
14.	Some of the Ooze of the Atlantic floor, magnified	124
15.	Iceberg at Sea	130
16.	The Birthplace of the Arctic Icebergs	132
17.	Scene among the disrupted ice of the frozen Arctic Sea . .	134
18.	The Ice-foot of Greenland	136
19.	Diagram illustrating the origin of the tides	149
20.	Diagram showing the relation of the beach to the lines of high and low water	152
21.	View of the sea-cliffs south of the river Tyne (magnesian limestone worn away by the waves, and the isolated fragments left) . .	159
22.	Steep shore descending abruptly into deep water	167
23.	Low shore shelving into shallow water	167
24.	Section of one side of a Continent	175
25.	Section to show the formation of soil and subsoil from rotting away of rock underneath	183
26.	Bedded arrangement of rocks	185
27.	Piece of shale containing portion of a fossil fern	186
28.	Piece of limestone, showing how the stone is made up of animal remains	187
29.	Some of the grains of a piece of chalk	188
30.	A piece of granite, showing the composition of a Crystalline Rock	188
31.	View of the geysers of Iceland. Great Geyser in eruption . .	193
32.	Plan of the Peak of Teneriffe, showing the large crater, partly effaced, and smaller craters with lava currents issuing from them	198
33.	View of a Street in Pompeii	199
34.	View of Vesuvius as seen from Naples during the eruption of 1872	201
35.	Mount Vesuvius as seen from the sea, with the remaining part of the old crater of Somma behind	202

LIST OF ILLUSTRATIONS.

FIG.		PAGE
36.	Houses surrounded and partly demolished by the lava of Vesuvius, 1872	204
37.	View of the extinct volcanoes of Central France, taken from the Puy de Pariou	208
38.	House rent by earthquake (Mallet)	215
39.	Diagram-section to illustrate the propagation of an earthquake-wave and the mode of calculating the depth of its origin	215
40.	View of an old sea-terrace or raised beach, with sea-worn caves on its inner margin	217
41.	Raised sea-terraces of the Alten Fjord, Norway	218
42.	Section of an island with a fringing coral reef	219
43.	Section of an island with a Barrier reef	219
44.	Section of an Atoll or coral island built over submerged land	220
45.	View of an Atoll or coral island	220
46.	Section across a valley to show how the simplest kinds of springs arise	228
47.	Section to show how deep-seated springs arise	229
48.	Section to show the intricate underground drainage which issues in a deep-seated spring	230
49.	Section to the position of the water-bearing rocks below the clay at London	231
50.	Section across the cliff and landslip of Antrim	242
51.	View of part of the cliffs and landslip of Antrim	242
52.	Delta of the Nile	246
53.	Windings of the Mississippi	249
54.	Part of the Island of Lewis, illustrating the abundant lakes of the north-west of Europe	259
55.	Section of a lake-basin excavated in solid rock	260
56.	Section of a lake-basin lying in a hollow of superficial detritus	261
57.	Section of a lake dammed up by a barrier of earth or gravel	261
58.	Prints made in soft mud or moist sand by rain-drops	273
59.	Earth pillars of the Tyrol	274
60.	View of ravines cut by streams out of a table-land	277
61.	Cascade and pot-holes of a water-course	278
62.	Section of a waterfall and ravine	279
63.	Terraces of gravel, sand, and mud, left by a river	287
64.	Delta of the Mississippi	292
65.	Snow-field and glaciers of Holands Fjord, Arctic Norway	301
66.	View of a glacier, with its lines of rubbish (moraines) and the river which escapes from its end	302

LIST OF ILLUSTRATIONS.

FIG.		PAGE
67.	Plan of the Mer-de-Glace of Chamouni and its tributary glaciers, showing the way in which lateral moraines become medial	305
68.	Glacier table—a pillar of ice supporting a block of stone	308
69.	The Pierre-à-Boat, near Neufchâtel	309
70.	Glacier descending to the sea. Head of Jokuls Fjord, Arctic Norway	310
71.	Stone polished and striated under glacier-ice	313
72.	Quarry in flat stratified rocks	316
73.	Section across a mountain-chain to show how the level rocks of the plains are bent and inverted along the flanks of the mountain, while the lowest and oldest rocks are made to form the central and highest point of the chain	318
74.	Section across a mountain-chain, showing two successive periods of uplift	319
75.	Section across a mountain-chain showing three successive periods of uplift	320
76.	Scene on the Coast of Caithness	322
77.	Portion of the west front of Salisbury Crag, Edinburgh	323
78.	Vertical distribution of climate on mountains	343

ELEMENTARY LESSONS

IN

PHYSICAL GEOGRAPHY

INTRODUCTION.

1. At night, when the sky is clear, the largest stars seem to stand out in front, with others less in size and brightness crowding behind them. As we gaze into these depths, still remoter and feebler twinkling points appear, until at last our eyes can no longer shape out any distinct specks of light. Such a sight impresses our minds, as nothing else can do so vividly, with the vastness of the Universe. We feel how comparatively small must, after all, be the distance which we can see into that "star-dust" which has been sown through the regions of space. And even when, with the aid of a good telescope, we return to these same skies, it is to find more cause than ever to acknowledge how immeasurably vast is that part of the Universe which man can thus explore; but at the same time, to meet again with a vague limit, beyond which we cannot see, not because we have reached the utmost verge of creation, but because our instruments can carry our vision no farther. Far beyond that limit, it may be that the regions of space contain other stars and systems, though too remote ever to be brought into

view even by the most powerful telescopes which human skill could construct.

Astronomers have calculated the distance of some of the largest and nearest stars. But their figures, expressing sums of many millions of miles, are too vast to carry any definite ideas to our minds. When we reflect that each of those stars, from the brightest to the faintest twinkling point, is really a sun, many of them, no doubt, far vaster in size than our sun, but dwindled into such seeming feebleness by reason of their inconceivable distance, we cannot but feel what a little speck of dust, in comparison, must this dwelling-place of ours really be which we call the Earth.

2. It is useful to get this comparative insignificance of the Earth firmly realised by our minds. And in no way can this be done so well as by watching the starry sky, and learning what has been discovered regarding the motions, sizes, and distances of the heavenly bodies. What, then, is the Earth in relation to these bodies? Has it always been in the same condition as now, or has it perhaps passed through long ages of change and progress? Mankind has had a long and varied history. May not the Earth itself have had one also? If so, can we learn anything about the story of the Earth?

3. Again, when, on the other hand, we look upon the face of the Earth by day, how boundless and varied it seems! From the district in which we may chance to live, we can pass in thought to the country at large, then to other countries, and then to the idea of the whole wide globe, with its continents and oceans, its mountains, valleys, and plains, and all that wonderful diversity of form and colour which makes its surface so unceasingly beautiful.

4. This variety is everywhere associated with life and movement. Consider, for instance, the unvarying succession of day and night; the orderly march of the seasons; the constant or fitful blowing of the winds; the

regular circling of the ocean tides; the ceaseless flow of rivers; the manifold growth and activity of plant and animal life! Surely it was no strange thought when men in old times pictured this world as a living being. And even though we cannot look on the earth as a living thing in the sense in which a plant or animal is so called, yet in view of all that multitudinous movement which is ever in progress upon its surface, and on which, indeed, we know that our own existence depends, there is evidently another sense in which we may speak of the life of the Earth.

5. Now this Life of the Earth is the central thought which runs through all that branch of science termed Physical Geography.[1] The word geography, as ordinarily used, means a description of the surface of the earth, including its natural sub-divisions, such as continents and oceans, together with its artificial or political sub-divisions, such as countries and kingdoms. But Physical Geography is not a mere description of the parts of the earth. It takes little heed of the political boundaries except in so far as they mark the limits of different races of men. Nor does it confine itself to a mere enumeration of the different features of the surface. It tries to gather together what is known regarding the Earth as a heavenly body, its constitution, and probable history. In describing the parts of the earth—air, land, and sea—it ever seeks so to place them before our minds as to make us realise, not only what they are in themselves, but how they affect each other, and what part each plays in the general system of our globe. Thus Physical Geography endeavours to present a vivid picture of the mechanism of that wonderfully complex and harmonious world in which we live.

6. Many of the facts with which this branch of science deals are familiar to every one. With nothing, for exam-

---

[1] This term as here used is synonymous with Physiography, which has been proposed in its stead.

ple, have we better acquaintance than with the air which surrounds us. We have breathed it every moment of our lives; we know it at rest as well as in storms; we have watched the growth in it of mists and clouds, as well as the fall of dew and rain. At first there might seem hardly anything more perfectly known to us, and therefore regarding which we should have so little new to learn. But Physical Geography, taking up the subject as one great whole, strives to show how all the different conditions and changes of the air are connected together, to explain their causes, and to point out the essential part which the air takes in the great movements that affect both sea and land. One of the great advantages of this study lies in the very commonplace character of the phenomena of which it treats. We cannot go anywhere without meeting with illustrations of some of the great lessons which it enforces. When we have once practically learnt these lessons, therefore, we carry an additional source of pleasure in every walk and journey. We get into the habit of using our eyes, and noting an endless variety of things that would otherwise have been passed by unseen, and this habit of observation, besides the pleasure its exercise brings with it, will be found of no slight service in the ordinary affairs of life.

7. In completing its picture of the Life of the Earth, Physical Geography must necessarily draw its illustrations from all parts of the world. It thus brings before us phenomena of which we may have no experience in our own country, and which, very probably, we may never have any opportunity of seeing for ourselves. In the study of it we journey, so to speak, all over the world, and learn in a short time far more about the world than we should be able to do from reading a few ordinary books of travel. Indeed, a good treatise on Physical Geography may be regarded as a condensed and well-arranged book of travels in all countries, with this dis-

tinction, that although it has no personal adventures to describe, it enables us to understand how one region of the earth differs from another, and it explains these differences by connecting them all together and showing their relation to the general principles or laws on which they depend. So that, although a man may never have been in India or Africa, or the Arctic regions, he may, from the study of Physical Geography, have a far better notion of the general features of these countries, and why they differ from each other so much in climate, than many other men who have travelled to, or even have lived for years in them. It is a matter of no little encouragement for all of us to know that the more we watch what takes place around us in our own country, and the more thoroughly we understand it, the more easily do we realise what goes on in other and distant parts of the world.

8. By taking up the study of Physical Geography, not merely as a subject to be learnt from books, but as a practical pursuit to be followed out by our own observations, as opportunity offers in the course of our daily occupations, we make most progress in it and get the largest amount of pleasure out of it. This is the spirit in which the following chapters are written. They are not meant merely to describe the different parts of the earth in such a way that these may most easily be learnt by heart, but rather to incite the learner to use his own eyes, and to examine, compare, and contrast what he sees to take place from day to day. They are so arranged as to begin, where practicable in each sub-division, with our common knowledge. Then they point out what can be ascertained on the subject by our own simple observation and experiment. Lastly, they present such further information as may be acquired from the observations and travels of men who have given much time and thought to the collection and investigation of the facts. In the case of the Air, for example, starting from what each one of us knows by everyday experience, we proceed to con-

sider what we can ourselves easily find out about the air, and from this basis of knowledge we follow what has been still further discovered by prolonged investigation in all parts of the world.

9. At the outset it may be well to group together in due order the different subjects which, coming within the scope of Physical Geography, will have to be attended to in these Lessons.

10. First of all we shall consider what the Earth is, as a heavenly body, how it is related to other heavenly bodies, and specially to the Sun, as the source of light and heat.

11. Secondly, looking at the Earth in itself, we find it wrapped round with an outer envelope of Air, which will next deserve our attention. What this envelope consists of, and the part it takes in the phenomena of the Earth's surface, furnish materials for much interesting inquiry.

12. Thirdly, beneath the surrounding shell of air, and covering the greater part of the surface of the solid globe, lies that vast expanse of water known as the Ocean. We shall follow its tides and currents, and trace the vapour which, ascending from its surface, is carried through the air until it falls as rain and snow upon the land, whence it is borne by rivers back again into the ocean. The wonderful beauty and high importance of this circulation will claim our attention.

13. Fourthly, the solid Land will be considered, with its continents and islands, its mountains and valleys, its earthquakes and volcanoes. The evidences of continual change on the surface of the land will lead us back to the action of the air and of water, and we shall find how largely the forms even of "the everlasting hills" have been determined by that action.

14. Fifthly, we shall inquire what Climate is, how different kinds of climate are distributed over the globe, and whether any causes can be assigned to account for such differences. This will lead us to note that as there is a geographical distribution of climates, so likewise is there one of plants and animals. Each great region of the earth's surface which has a peculiar climate of its own, has also its own distinguishing assemblage of peculiar plants and animals. Even in the way in which the races of man are grouped over the earth there is evidence of the same connection between the distribution of climates and of life. Thus the geographical distribution of Life over the earth's surface will form the concluding part of these Lessons.

CHAPTER I.

THE EARTH AS A PLANET.

Lesson I.—*The Earth's Form.*

1. What then is this Earth on which we live? The answer to this question must be chiefly given by astronomy, but there are some familiar features in the earth itself which, if rightly considered, help to make the answer more vividly realised. Evidently there is much advantage in obtaining a clear idea of what the earth is as a whole, before we begin to trace the events which take place upon its surface.

2. In the first place, then, the earth, like the sun and moon, is a globe. This may be proved in various ways.

(1) If we set sail from England, and steer westward without turning back, we shall eventually find ourselves in England again. This is called **Circumnavigating the world.** It would not be practicable unless the earth were really a globe.

(2) Standing by the margin of the sea, and looking seawards, we observe that a ship which is moving away from us, by and by seems to sink into or under the sea. (Fig. 1.) The hull first disappears, and then by degrees the sails. So, on the other hand, when a vessel approaches us, we can first, with a telescope, make out the tops of the sails and masts just peeping above the distant surface of the ocean. Little by little the sails rise as it were out of the water, until at last the hull and the whole vessel come into sight. This could not happen

unless the surface of the sea instead of being flat were really curved, that is, part of the curved surface of a globe.

(3) If the surface of the earth were flat, as men once supposed it to be, the sun would rise at the same time at all places on the earth. But this is not the case; sunrise takes place later or earlier as we travel west or east. Again, on ascending a mountain we should see the whole of the earth's surface if that surface were really a plain. But the extent to which we see depends on the height to

Fig. 1.—Curvature of the Earth's Surface, as shown by ships at sea.

which we climb. This shows that the earth must be a globe.

(4) When the earth comes directly between the sun and moon, so as to cut off the light of the former from the latter, the moon is said to be eclipsed. Now, the earth's shadow as it creeps over the moon's surface, is seen to be circular, and hence the real form of our planet must be that of a globe.

3. Were it possible for us to quit the earth and look back upon it from a distance of a few millions of miles, it would appear as a large bright moon, hanging in space, with the stars twinkling on all sides of it. Its surface would shine like that of the moon, with the light falling on it from the sun, and would be seen to be marked by

irregular bright and dark strips and patches. There would be two specially bright parts on opposite sides of the earth's body, marking the great regions of snow and ice at the north and south poles. Between these two areas certain bright irregular bands would indicate the

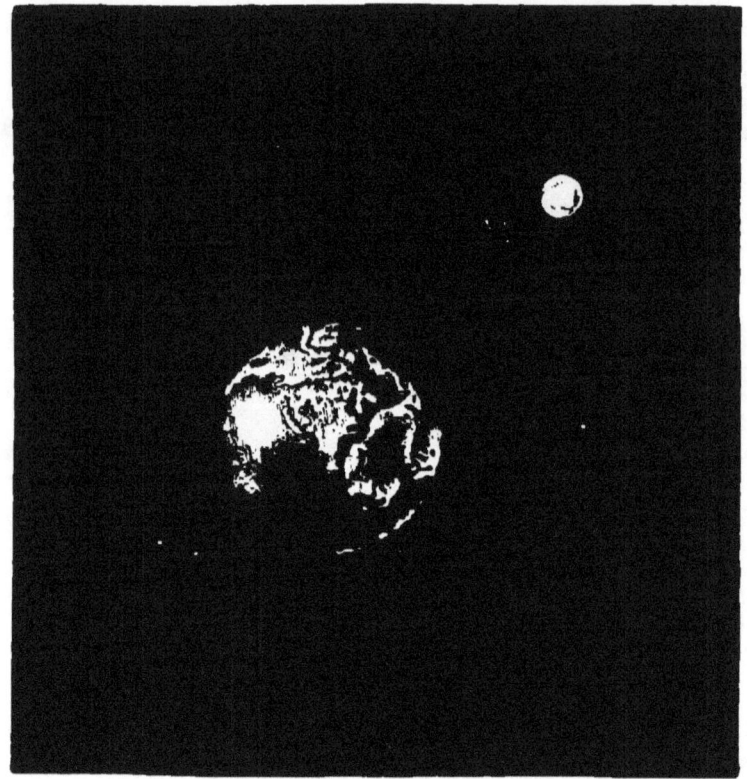

FIG. 2.—The Earth and Moon as seen from space.

position of the dry land, while the intervening darker portions would show the extent of the seas (Fig. 2). If we could transport ourselves still farther away; if, for instance, we could get as far off as the sun, that is to say more than ninety-two millions of miles from where we are living at this moment, our earth would be seen shining merely as a bright star. And were it possible for us to reach one of the nearest fixed stars, our globe would be no longer visible, while the sun itself, if seen at all, would appear merely as a twinkling star.

4. But the earth is not a perfectly regular sphere. In particular it is somewhat flattened on two opposite sides, so as to resemble an orange in shape. How this is known and measured will be seen from Lesson IV. A line from the middle of one of these flattened sides passing through the centre of the planet to the middle of the opposite flattened side, is called the **Axis**, and the points at which it reaches the surface are known respectively as the **North Pole** and **South Pole**. A line drawn midway between the poles round the bulging part of the earth is termed the **Equator**.

LESSON II.—*The Earth's Motions.*

1. So long as men in early times looked upon the earth as a great plain placed in the centre of the universe, with sun, moon, and stars moving over and under it every day and night, they could not for a moment imagine that the Earth itself is moving. But the truth in this matter has now been so long known, that we have no difficulty in seeing how the apparent movements of the sun and stars are explained by the real movements of the earth. Of these movements the two chief are called **Rotation** and **Revolution**.

2. **Rotation.**—The most obvious movement of our earth is its rotation on its axis, to which the succession of day and night is due. A complete rotation is performed in about twenty-four hours. During that time each part of the earth's surface is alternately presented to, and turned away from, the sun. As the sun rises in the east to give us day, and appears to travel westward across the sky, the real motion of the earth must be in the reverse direction, or from west to east.

3. It is evident that as the earth rotates, the rate of motion and the distance travelled will vary for different parts of the surface according to their nearness to or distance from the axis. When a cart-wheel, for example, is made to spin round on its axle, a mark made on its rim must move faster and describe a much larger circle than one made near the axle. Places situated on the earth's equator must have the maximum velocity and travel over the greatest distance, while at either pole the velocity must be *nil*. The circle described by a place on the equator is indeed the whole circumference of the globe. So that when the number of miles in the circumference is divided by the number of hours required for a complete rotation, we obtain in figures the rate of motion. At the equator this rate is found to be upwards of 500 yards in a second; it gradually diminishes to the poles.

4. Now if the earth is rotating with such speed, why are we not jerked off its surface? When a stone is thrown into the air, why does it not immediately rush away into space instead of falling back again to the ground? Because what is called the **earth's attraction** is far stronger than this tendency to fly off. Everything in and on the surface is pulled down towards the centre of the earth by this attracting force, which is called **gravity**. But it will be seen that the influence of rotation is nevertheless unmistakably marked upon the great atmospheric currents, since these have their direction modified by it (Lesson XI. Art. 13).

5. **Revolution.**—The earth revolves round the sun and takes about 365 days to perform the circuit. What we call a year is simply the length of time taken by the earth to make one complete revolution round the sun. The path which it follows, called its **orbit**, has been ascertained not to be a perfect circle but an ellipse, so that our globe is at one part of its course nearer to the sun than at another, its mean distance being computed

at 92,800,000 miles. From this number and from the time taken for a single revolution it is easy to find the average rate at which our globe must rush along its orbit. That rate is about 68,000 miles in an hour.

6. There are certain circumstances connected with the movement of revolution which serve to explain why the days and nights throughout the year are not all of the same length, and why, instead of perpetual summer or winter, there is the regular succession of the Seasons. If the axis of the earth were perpendicular to the plane of the orbit, that is, if it moved along in a perfectly upright position, there would be equal day and night all the year round all over the globe. But it is really inclined to the path of the planet round the sun at an angle of about $66\frac{1}{2}°$ and remains always parallel to itself, that is, always points to the same star. In the summer of the northern half of the globe, the North Pole is turned towards the sun, in winter it is turned away from it, the South Pole being at the same times in the opposite positions. Hence while midway between the two poles the day and night remain each of twelve hours' duration, they depart more and more from this uniformity towards the poles, until there we have a day lasting for one half of the year, and a night lasting for the other half.

7. About the 22d of March, and again about the 22d of September, the earth is so placed in its orbit that the sun is exactly vertical over the equator. The line between the sun-lit half and the dark half passes through the poles. At these times, therefore, day and night are each of twelve hours' duration over the whole globe; and these points in the year are accordingly called **Equinoxes**, that is, "equal nights." After the March equinox, the northern parts of the globe, as the earth moves round in its orbit, come more and more into the sunlight; in other words, the days get longer and longer until, at the North Pole and for a certain space all round it, within what is called the **Arctic Circle** (see Plate

I.), the sun does not set at midsummer at all, and there is continuous daylight. The farther north we go the longer are the days in June, so that a summer-day in England differs very much in length from one in India, and one in the north of Scandinavia from one in England.

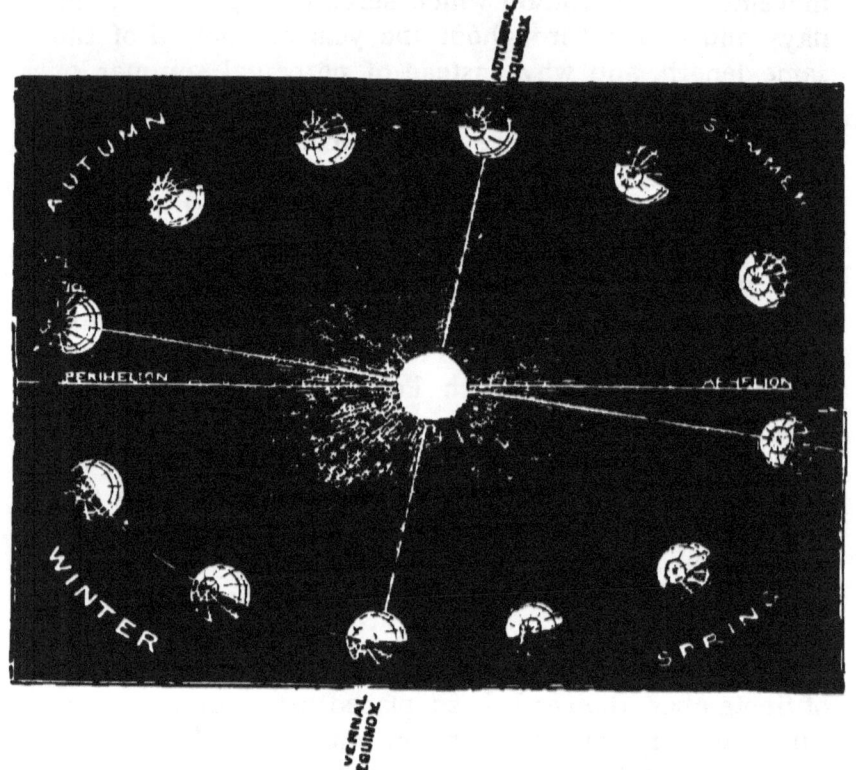

Fig. 3.—The Earth's path round the Sun.

On the other hand, when the earth has reached the September equinox, the North Pole begins to point away from the sun, and the days in the northern part of the earth begin to grow shorter, till in midwinter, at the pole and over the ground within the Arctic Circle the sun does not rise, and there is continued night. The farther north we go in December, therefore, the longer are the nights; so that while a summer-day in London is not so long as a summer-day in the north of Scotland, a winter-day is considerably longer.

8. It is evident that when the North Pole is enjoying perpetual daylight, the South Pole must be in continuous night. It is only at the two equinoxes that their day and night are equal. As the days about the North Pole shorten, those at the opposite pole lengthen, and then the reverse takes place, and so on from year to year.

9. The alternation of the Seasons depends, in like manner, upon the inclination of the earth's axis in its yearly orbit. At one part of the year, people in Europe and North America see the sun comparatively low in the sky: his rays are then but feeble; the climate is cold, and then comes the familiar sight of the frost and snow of *winter*. Six months later the sun appears much higher in the sky; that is, more directly overhead: his rays are then warm—even scorching, and we find ourselves in the midst of *summer*.

10. In reality, however, it is not the sun which has changed place, but our planet, which has reached different parts of its orbit, and consequently has presented itself in different positions towards the sun. In summer the days in the **Northern Hemisphere** (as the half of the globe between the equator and the North Pole is called) are long, because the North Pole is turned towards the sun. That hemisphere is consequently warmed. The rays of the sun come more directly down upon it; the long days allow much more heat to be received from the sun, and the short nights permit much less heat to be given off from the earth than in winter. While this state of things in the Northern Hemisphere brings summer there, precisely the opposite effects are taking place in the Southern Hemisphere. There, in proportion as the days lengthen and the warmth increases in the North, the days shorten and the heat decreases, until at the midsummer of the Northern half of the globe, the people at

the **Antipodes**, that is, on the opposite side, or Southern Hemisphere, are in their midwinter. Christmas, which in Europe and North America is always associated with the frost and cold of midwinter, happens at the midsummer of countries in the Southern Hemisphere, like Australia and New Zealand.

11. It is evident that the contrast between summer and winter must be most marked in the region round each pole, where indeed the year may be looked upon as consisting of only two seasons, one of daylight and summer, the other of night and winter. In the equatorial belt, too, that is on either side of the equator as far as the lines called the **tropics**, there is often a striking difference between the seasons, according to the position of the sun in the sky. At each equinox the sun appears vertical over the equator. From March to June he seems to be travelling northward until about the 22d of the latter month, when he shines vertically over a line called the **Tropic of Cancer**, $23\frac{1}{2}°$ north of the equator. He then turns southward, is again vertical above the equator at the September equinox, after which he travels southward until about the 22d of December, when he shines directly overhead along the line known as the **Tropic of Capricorn**, $23\frac{1}{2}°$ south of the equator. The word tropic means "a turning" and is applied because the sun after appearing to travel away from the equator begins to turn at these two limits on each side. The sun seems to be constantly travelling to and fro across the sky between the two lines of the tropics. This apparent movement of the sun is of course due to the real movement of the earth, and the limit within which the sun travels north and south of the equator is fixed by the angle of inclination of the earth's axis ($66\frac{1}{2}°$).

12. In the belt between the tropics the sun appears vertical in the sky twice in each year. With each of these periods comes a rainy season, and between them a dry and hot season. So that if no other influences came into operation the equatorial regions would have two wet and two dry periods in each year. Owing, however, to the way in which the wind currents are interfered with by the masses of high land, it is in the oceanic regions that this arrangement of the seasons is most completely carried out.

LESSON III.—*The Earth and the Sun.*

1. Long before astronomy or any science had arisen, the early races of man, seeing how greatly the earth is dependent on the sun, worshipped that luminary as the great parent of all the light, and heat, and life of the world. And surely of all the forms of idolatry none is so natural as this. Before they knew what the sun really is, men might well be content to prostrate themselves before it as a great and good Being, who when he rose at dawn, lighted up and warmed the world, and when he sank at dusk, left it in darkness and chill.

2. But how or why is it that our light and heat come from the sun? We know that the earth revolves round the sun, but why should this be? Is it possible to know anything as to the real relation of the earth to the great luminary?

3. Let us try to find an answer to these questions by putting down some of the facts which have been discovered about the earth and the sun. In doing this we shall see how step by step the evidence grows, that the dependence of the earth upon the sun is so close because in reality the two bodies originally formed part of one great rotating mass of matter.

4. To begin with the earth. When a deep bore is made into the ground in search of water, as has been done round London and Paris, sometimes to depths of more than a quarter of a mile, the water which rises from so great a distance is found to be warm.

In like manner the air of deep mines is often so hot that the miners have to work with hardly any clothes on them. In many parts of the world springs of hot and even boiling water rise to the surface, and have been doing so for hundreds of years without becoming perceptibly cooler. Again, in all quarters of the globe volcanoes, or burning mountains, are found from which steam, hot gases, and molten rock are from time to time given out, often with prodigious force and in enormous quantity (see Lesson XXII.).

5. From all these facts which have been observed throughout the world, the inference has naturally been drawn, that the inside of the earth must be intensely hot. And since the depth of even the deepest bores or mines is not, in comparison to the whole mass of the earth, so much as the thickness of varnish is to an ordinary school globe, it would seem that the earth must be, on the whole, a hot globe with a cool outer skin or crust.

6. Since its surface is cold, the earth cannot give out any light of its own. And yet it undoubtedly shines in the sky, as the moon does, because like that orb it receives and reflects light from the sun.

7. Passing from the earth to the nearest of the heavenly bodies—the moon, which as the earth's satellite or attendant, revolves round our globe, while the latter is revolving round the sun, we find some wonderful evidence that the earth is not the only orb that shines with light borrowed from the sun, and though cool on the surface, bears traces of the influence of internal heat.

8. The nature of this evidence may be followed by comparing the two drawings, Figs. 4 and 5. In Fig. 4 a sketch or plan is given of part of the neighbourhood of Naples. The numerous detached conical hills are volcanoes, either now active as Vesuvius is, or cold and silent like the group of cones to the west of Naples. The circular pits placed on the tops of these hills are the "craters" or vents from which dust, cinders, steam, and

molten rock, have at different times been copiously discharged. In Lesson XXII. some account will be given of volcanoes, and it will there be shown how the streams of molten rock now and then run down the outside of the cone, while vast clouds of steam and hot dust, cast up with great violence, bear further witness to the intense heat of the interior.

Fig. 4.—Plan of Volcanic Hills and Craters in the Bay of Naples.

9. Now, with this little piece of the earth's surface compare the drawing in Fig. 5, which shows a portion of the surface of the moon. We observe in it a series of large "craters," with steep walls in the inside, and numerous smaller ones, sometimes in close-set rows. Several of the larger craters overlap each other, and in some cases their interior is crowded with the smaller ones, or the latter have broken through the ridges separating the great basins. So close is the resemblance of this scene to parts of the earth's surface, like that shown in Fig. 4, that there seems no reason why we should not call these conical elevations on the moon volcanoes, and look on their well-marked "craters" as really the open-

ings whence molten rock and other hot materials have been thrown out, as in terrestrial volcanoes. So far as can be seen they abound over most of the moon's surface. Indeed, the lunar volcanoes have been measured by astronomers, and found to be far more numerous, and more gigantic than the terrestrial ones. We may fairly infer that they mark a much more intense kind of volcanic action than anything we know of on the earth.

Fig. 5.—A part of the Surface of the Moon, showing Volcanic Craters.

So that though the chill surface of the moon now gives out no light of its own, but only throws back to us that which strikes upon it from the sun, the interior of that globe must once have been intensely hot, with a hardening crust through which the internal molten liquid was poured by those hundreds of craters and rifts.

10. No telescope has yet detected any actual volcanic eruption going on in the moon. It would seem, indeed, that the once prodigious volcanic action there has spent itself, and that the inside of the moon, though perhaps still warmer than the outside, has not heat enough left

for further outbursts. Evidently the moon has cooled very greatly since the time when its craters were active volcanoes.

11. The earth must be cooling likewise. The mere difference of temperature between its surface and interior indicates this fact. The outer shell, or crust, would not be kept colder than the inner mass unless there were a continual loss of heat from the earth into space. Heat is steadily passing outwards through the crust and surface, but it does not make the ground on which we walk sensibly warmer, because as fast as it reaches the surface it is radiated into space.

12. Thus we see that the earth cannot have been always as it is now. A million of years ago it must have had considerably more heat than it possesses to-day. A hundred millions of years ago it was, perhaps, in the condition of a globe of molten matter, with no land or sea, and of course with no life of any kind upon its surface. The present flattening of its form at the poles is just the kind of shape which such a liquid globe would necessarily assume under the influence of rotation, and probably the permanent flattening dates from that time. Earlier still the earth may have existed merely in the state of vapour.

13. That this has really been its history has been thought by many philosophers to be in the highest degree probable, when, turning from the evidence which the earth itself presents, attention is given to that furnished by the sun. That the sun is hot has long been familiar knowledge, but only in recent years has any adequate notion been formed of how intense its heat actually is. One fundamental difference between the sun on the one hand, and the earth and moon on the other, is that its light is given out by itself, while theirs is only borrowed sun-light. When sun-light is examined by means of the spectroscope, it is found to indicate a temperature in the sun so enormously

high that nothing can exist there except in the form of gas or vapour. The light and heat which we receive from the sun proceed from glowing vapours, among which have been detected those of some of the metals found on the earth only as solid bodies very difficult to melt. Probably most or all of the simple substances composing the earth exist also in the sun, but in the form of vapour. Were our globe thrown into the sun it would immediately be dissipated into the same glowing vapour.

14. Observation of the sun's surface has shown that spots which appear there are carried steadily round from west to east. This points to a movement of rotation, which, however, is slower than that of the earth. The sun, or at least its outer luminous envelop which we see, takes about twenty-five of our days to turn round once on its axis; but the movement is in the same direction as that of the earth.

15. Besides the earth and its attendant moon, a small number of other heavenly bodies has been found to revolve round the sun. Noting carefully the positions of the stars in the sky at night, we find that, though the whole sky seems to travel slowly westward, each star keeps the same place with reference to the other stars. But we may observe a few which look at first like stars, but which, when watched attentively from time to time, may be seen to shift their place and to travel across the other stars. These were called by the ancients, **planets**, that is, wanderers. They are now known to be really revolving round our sun at different distances and in various periods of time. So far as can be judged from what the telescope reveals, they closely resemble the earth in many points. They rotate on their axes. Some of them have a group of attendant moons. In some there are indications of an atmosphere with clouds and air-currents, and in one of them, called Mars, there appears to be a cap of snow and ice at each pole, as in

the earth. Some are much larger, others much smaller than the earth; some are nearer to the sun than we, others greatly farther away. This whole series of planets (including, of course, the earth), which revolves round the sun as its centre, is known as the **Solar System**.

16. Now let us sum up all these facts, and see how they bear upon the present condition and probable history of our own globe. In the centre of the solar system stands the sun—an enormous globe of incandescent gas and vapour, rotating on its axis, and sending forth heat and light far and wide through space. Round this central luminary, and depending on it for light and heat, a number of planets revolve in the same general plane, sometimes with minor planets or satellites revolving round them, as the moon does round the earth. The planets and their attendant satellites or moons likewise rotate on their axes. The earth is one of the planets. Its present condition points to its once having been hotter than it now is, and probably to its existence in a liquid or even gaseous state.

17. The **nebular theory**, which has been proposed to connect and explain these facts, maintains that at first the whole solar system existed only in the state of vapour, like one of the faint cloud-like **nebulæ** disclosed by the telescope among the stars; that this nebula gradually condensed and threw off from its hot mass successive portions, which, by subsequent condensation and cooling, became planets, and that the present sun is the remaining incandescent, but still slowly condensing and cooling nucleus of the whole, round which the various detached portions continue to revolve.

18. This theory, which the discoveries of modern science go far to confirm, gives an intelligible reason for the intimate way in which the earth and the other planets are related to the sun. It furnishes a satisfactory explanation of the fact that the inside of the

earth is still hot, though slowly cooling. This internal heat, manifested in every deep bore and mine, as well as in hot springs and volcanoes, appears thus as a residue of the original heat of the great nebula out of which the planets and the sun have been condensed.

LESSON IV.—*Measurement and Mapping of the Earth's Surface.*

1. It would not have been possible for geography to make much progress without some means of accurately ascertaining the positions of places on the earth's surface. On a small scale we can measure with great exactness by means of carefully-constructed chains or rods, how far one place is distant from another. But evidently this laborious method could have been of little use in laying down the great features of the earth's surface—continents, islands, oceans, and the rest—or in discovering the actual dimensions of the planet itself. It was needful to obtain some other method which could be easily used, and would give the means of determining with precision the position of every part of the surface of the globe. Such a method is supplied by observing the position of the sun and different stars.

2. If we note the height of the sun in the sky at noon, we shall find that an hour after that time he seems to have moved a certain distance westward. In the course of another hour he will have traversed another similar space, and so on hour after hour till nightfall. Next morning, if we again watch his progress, we detect a similar advance hour by hour, until at noon he once more stands in the same position as at noon of the day before. In this interval the earth has made one entire rotation.

3. Every circle is divided into 360 equal parts, or **degrees**, and as we travel round our terrestrial circle in twenty-four hours, we must pass over fifteen degrees

A	C		F
O	C	E	A

PLATE I.

FIG. 1.

every hour. Suppose that some place, say the Greenwich Observatory, is fixed on from which to count the degrees. It is clear that all places lying to the east of Greenwich have their noon earlier, and all places to the west have it later than at Greenwich itself. As the sun takes an hour to travel 15° of the circle, any place which has its noon exactly one hour after noon at Greenwich must lie 15° to the west.

4. To determine, therefore, how far we are to the east or west of Greenwich, we may try to find out the difference between Greenwich time and the local time at the place where we may be. This has been done for short distances by flashing mirrors, exploding gunpowder, or in any other way communicating an instantaneous signal from one point to another. The most efficacious method is by electric telegraph. But for long journeys, especially sea-voyages, where no such communication is possible, carefully-constructed clocks called chronometers are used, which show Greenwich time. By comparing the local time, as fixed by taking the sun's position, with the time shown in the chronometer, it is easily seen how far a place lies to the east or west of Greenwich.

5. But as even the most accurately finished clock is apt to gain or lose, there is still another and more reliable, though less convenient kind of observation—that of determining the position in the sky of the moon or some of the planets. The places which these bodies will have with reference to each other and to the fixed stars at any moment is calculated for a long time beforehand at the Greenwich Observatory, and tables showing these positions are printed. By referring to these tables the traveller finds the precise instant of Greenwich time, for each position of the heavenly bodies; and the difference between that time and the time he observes at the place where he is shows him whether and how far he is to the east or west of Greenwich. This is called " **finding the Longitude**" of a place.

6. If each of the 360 degrees in the circumference of the globe were marked by a line passing from pole to pole, each line would be found to run due north and south. A little reflection will show us that every place on the earth's surface lying along the course of the same line must have its noon at the same moment. These lines, supposed to be traced on the earth's surface and actually placed upon maps, are termed **meridians** (Plate I. Fig. 1). Of course we may fix on any point from which to begin to count them. In England and in English-speaking countries they are reckoned from Greenwich Observatory. In France they are counted from the Observatory at Paris. But this is immaterial, though it would be a great convenience if all countries would agree to count from the same starting-point. There is reason to hope that this may before long be accomplished, and that Greenwich will be selected. The meridian line passing through Greenwich is marked by us as the zero or 0°. When it is twelve o'clock at Greenwich it is precisely the same time at every place through which the meridian of Greenwich passes. All the meridians lying west are West Longitude, those to the east are East Longitude, so that the two series meet on the other side of the globe, exactly opposite to Greenwich, at 180°, or half the number (360) into which the whole circle of the globe is divided. Having then fixed on our meridian as the starting-point, it is not difficult to see how the relative distances of places east or west of that meridian are determined and expressed. We say that a place which lies five degrees east or west from Greenwich is in 5° East or West longitude. Each degree is further divided into sixty minutes, and each minute into sixty seconds. Paris is situated two degrees, twenty minutes, and nine seconds east from Greenwich, which is abbreviated thus: 2° 20′ 9″ E.

7. One degree of longitude is equal to four minutes of time. When we travel eastwards our watches seem to be going slower at the rate of four minutes for every such degree, because we are getting into longitudes where the time is earlier than at home, while, on the other hand, when we travel westwards, our watches seem to be going too fast by the same intervals.

8. But this effect of difference of longitude is brought out in by far the most striking manner in the sending of messages by electric telegraph. Though two places may be thousands of miles apart, a word sent from one is almost instantaneously received at the other. When a clerk in London telegraphs at noon to Calcutta (which lies in 88° 30′ E. longitude) his words, though they go with the speed of lightning, arrive about six o'clock in the evening by the clock in India; or should he send a telegram at the same hour to New York, which is situated in 74° W. it will find the clocks there pointing to seven o'clock in the morning.

9. But the determination of the longitude of a place would not be enough. We must be able to tell whereabouts that place lies on its meridian. To do this is what is termed "**finding the Latitude.**" The first thing to strike us in this problem is the fact that we do not require to fix on any arbitrary point from which to begin our reckoning. The axis of the earth gives us two definite points at each pole, and midway between these lies the line of the equator. To determine the latitude of a place, therefore, we have to ascertain how far it lies to the north or south of the equator. Here again we must have recourse to the heavenly bodies.

10. If the axis of the earth were prolonged from the North Pole, it would reach a point in the heavens close to the pole-star, called the **Celestial Pole**, round which in the northern hemisphere the stars, by reason of the earth's rotation, seem to revolve. So that though we cannot measure from the North Pole itself we can determine our distance from it by observing how much the point of the heavens directly above us (that is, the

zenith) is removed from the pole-star or from the celestial pole.

In the same way if the line of the equator were prolonged it would traverse the heavens as a great circle. Knowing exactly its position in the sky, or the distance of any heavenly body from it, we observe how far our zenith is removed from it, and thereby learn our distance from the terrestrial equator.

11. Now the distance from either pole to the equator is exactly a quarter of a circle, or 90°. If we mark off the degrees by a series of lines on the surface of the earth, these circles will run round the globe parallel to the equator and to each other, but diminish in diameter as they advance to the pole. Such lines are called **parallels** of latitude (Plate I. Fig. 2). They are counted from the equator, which is the zero, or 0°; and each degree is divided into minutes and seconds like the degrees of longitude.

12. If we find by observation that a place is situated fifteen degrees north and another twenty degrees south of the equator, we say that the one is in Latitude 15° North, and the other is 20° South. Each pole is hence in latitude 90°. Since the figures expressing the degrees increase in amount as they recede from the equator, it has become common to speak of "**low latitudes**," that is, regions or places lying towards the equator, and "**high latitudes**," that is, tracts situated near either pole.

13. The degrees of latitude and longitude, then, are defined by two sets of lines, one running north and south, the other running east and west, by which the surface of the globe is supposed to be traversed as by a regular network. By means of these lines it is clear that we can fix exactly the **relative** positions of places on the earth's surface. But we must still determine what is the **absolute** length of one of those degrees in miles before we can ascertain the true distances of places and the real area which any

country, or continent, or sea covers. When this is done we have the materials for estimating the total bulk of our planet, and making a correct survey of its surface.

14. Since the lines expressing the meridians of longitude converge to each pole, their distance from each other must vary according to latitude; the parallels of latitude necessarily diminish in circumference as they recede from the equator. But were the globe a perfectly symmetrical sphere, the length of each meridian from equator to pole should be exactly alike. The accurate measurement, therefore, of the length of one degree of a meridian should give us the means of easily computing the sum of the whole, and thence of ascertaining the true size of the globe.

15. This **measurement of a degree of the meridian**, as it is called, has been made with great care in different parts of the world. In India the length of a degree was found to be 362,956 English feet, or rather more than $68\frac{3}{4}$ statute miles; a degree in Sweden measured 365,744 feet, or, in round numbers, about $69\frac{1}{4}$ miles. It has been found that, besides little irregularities which show that the shape of the planet is slightly distorted, there is a progressive increase in the length of the degrees towards the poles, as these two measurements in India and Sweden show. This could only take place by a flattening of the globe at the poles.

16. The sum of all the observations gives for the **polar diameter** of our globe a length of 7,899·17 statute miles, and for the mean **equatorial diameter** a length of 7,925·65 statute miles. As the difference is about $26\frac{1}{2}$ miles, each pole must be compressed to the extent of $13\frac{1}{4}$ miles.

It is not easy at once to grasp the full value of these figures. An express train travelling at the rate of thirty miles an hour would take about a month to go completely round the earth at the equator; and if it could go through the earth between the poles, it would take about eleven days to perform the journey. Upon the surface of a globe of such dimensions, the highest mountains and the deepest oceans are far less in proportion than the roughnesses upon the rind of an orange.

17. Astronomy has taught us not only the size of our own planet, but has computed the dimensions of the others, whence we learn the comparative place of the earth in the solar system. The planet Jupiter, for example, is 1400 times the size of the Earth. On the other hand, our planet is seventeen times larger than Mercury, and greatly larger than certain small bodies called Asteroids. Again, the Earth is neither nearest to nor farthest removed from the sun; its mean distance, as already stated, is computed to be nearly 93 millions of miles. But Mercury, at its greatest distance, is only about $44\tfrac{1}{2}$ millions of miles away from the sun, while Neptune revolves at the enormous mean distance of 2862 millions of miles. The sun itself, the great centre of all the movement of the solar system, would contain 1,400,000 globes as large as ours.

18. Having found a means of accurately fixing the position of any place on the earth's surface, and of determining the distance of places from each other, men would nevertheless have been able to carry in their minds but a vague notion of the general features of that vast and varied surface, had they not devised a way of putting down these features upon paper so as to show their relative positions and shapes. Such a delineation of the whole or part of the earth's surface is called a **Map**. The map of any country or continent may be seen to be crossed by two sets of lines, one of which, running from the top to the bottom of the paper, marks the degrees of longitude, while the other set, running from side to side, shows the parallels of latitude. It is usual in such maps to represent the ground as it might be supposed to appear could we ascend to a great height in the air and take in the whole area at one view, our heads being

turned to the north. Hence the top of the map is north, the right hand being the east side, and the left hand the west.

19. When maps of a country are wanted on a large scale to express the features of the ground in great

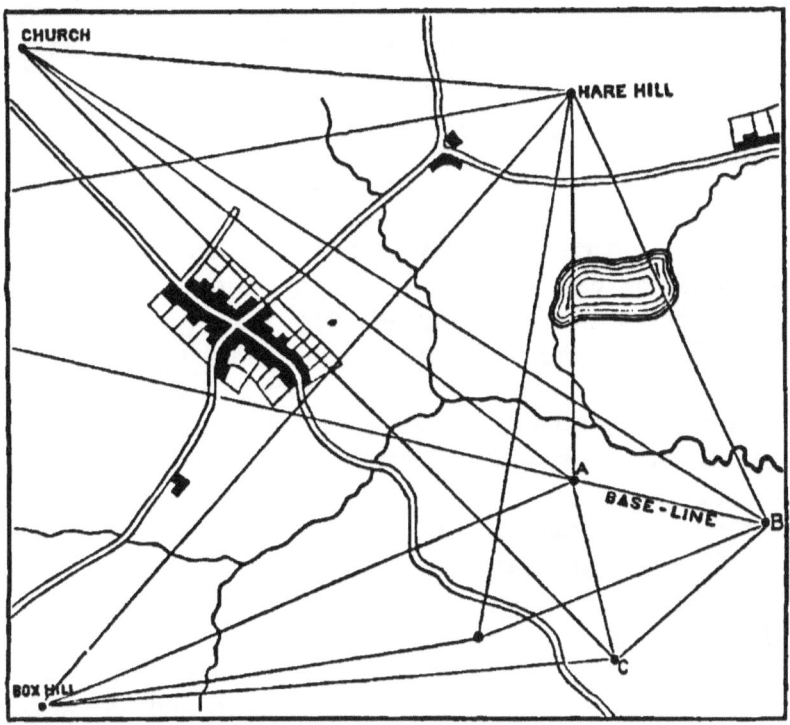

FIG. 6.—Measurement and mapping of a country by means of triangulation.

detail, another process of measurement called **triangulation** is used. First a **base-line**, a few miles in length, is accurately measured with rods or chains as from A to B in the figure. Then from A an observation is taken with an instrument called a theodolite, to the point C, which may be a hill-top or church tower, or any distinct object, and the angle C A B is carefully noted. A similar observation is taken at B to determine the angle

C B A. We can now construct our first triangle, and having found by actual measurement the length of its base, we find by calculation the length of its two sides, and consequently the exact position of C and its distance from A and B respectively. From such a measured base a system of triangles is observed all over the country, and the true positions of all the main landmarks are fixed by triangulation without the necessity of further actual measurement.

LESSON V.—*A General View of the Earth.*

1. We proceed now to take a first general view of the parts of the earth with which further acquaintance is to be made in the following lessons. Our progress will probably be found to be more easy and pleasant, if we can carry with us from the outset such a broad but clear notion of the earth as a whole, as will enable us to follow the necessary references to parts with which we may still be unfamiliar

2. Beneath the outer envelope of **Air**, with its winds, clouds, rain, and snow, the surface of the globe presents the two clearly marked, but very irregular divisions of **Sea** and **Land**. As the result of surveys and observations taken in all parts of the globe, it is ascertained that the sea covers nearly three-fourths, and the land rather more than one-fourth, of the whole surface of the planet; or, more exactly, there are 275 parts of water to 100 parts of land. The total area of the earth's surface is computed at 196,712,000 square statute miles, of which 144,712,000 are assigned to the ocean, and 52,000,000 to the land.

3. For the sake of aiding our conceptions of such statements as these, globes and maps are prepared, of which Plate I. may serve as an illustration. Each of the two large circles there represents one side of the

earth, and shows the manner in which sea and land are grouped. It will be observed that, while the amount of sea greatly exceeds that of land, the difference becomes more marked in the southern half or hemisphere, which is almost all water, while most of the land lies on the north side of the equator. But a school globe may be so placed, as to show nearly the whole of the land at one view, and the greater part of the sea at another. For this purpose, turn the globe so that the south-west of Britain appears directly in the centre of the globe as you look at it (Fig. 3, Plate I.). Britain is thus seen to lie in the centre of the habitable part of the earth. Now turn the globe so that the point on the other side, exactly opposite to the south-west of Britain, shall be the centre. The islands of New Zealand are then not far from that position, and all round them lies the water hemisphere, with comparatively few detached masses of land (Fig. 4, Plate I.).

4. While the great body of the land lies on the north side of the equator, it will be seen from Plate I. that it forms there two well-marked portions separated by two great tracts of sea. The larger of these portions includes Europe, Asia, and Africa, that is, all the regions which have been longest settled by a human population. Hence it is often spoken of as the Old World. The other portion embraces America, and is called the New World. As the Old World lies to the east and the other to the west, they are frequently referred to as the Eastern Hemisphere and Western Hemisphere.

5. One difference between the distribution of sea and land cannot fail to arrest attention at the outset. The land is much broken up. Even in the Eastern Hemisphere, where it presents the most compact mass, long arms and inlets of the sea intersect it, and cut off large portions from the rest. The sea, on the other hand, is evidently one continuous whole. Even when penetrating farthest into the land its most distant arms retain

their connection with the main body. A vessel may sail over every sea and into every far recess and inlet, without ever needing to be dragged over an intervening mass of land. But no coach or form of carriage could visit every part of the land, without requiring to be transported across intervening tracts of sea. There are no isolated areas of sea completely encircled by land, answering to the abundant portions of the land which, encircled entirely by sea, receive the name of islands.[1]

6. Though the sea is one continuous liquid mass it has been, for the sake of convenience in description, divided into different areas, termed **Oceans**. The limits of these are for the most part indicated by the position of the masses of land. Thus the largest, under the name of the Pacific Ocean, fills up the vast area between the western edge of America and the eastern margin of Asia and Australia. The line of the equator divides it into the North Pacific and the South Pacific Ocean. Another longer but much narrower belt of sea, intervenes between the eastern coasts of America and the western coasts of Europe and Africa; divided in the same way by the equator into the North Atlantic and the South Atlantic Ocean. Owing to the broad mass of land forming Europe and Asia, no corresponding belt of sea can run there from pole to pole. But south of that land lies the wide Indian Ocean, between Africa and Australia. It is usual to call all that part of the sea within the Arctic Circle the Arctic Ocean, and the corresponding tract in the southern hemisphere, the Southern or Antarctic Ocean.

7. Besides these main sub-divisions there are minor tracts of sea, more or less surrounded by land. Such names as sea, gulf, strait, channel, are given to them, according to their size or the shape of the land-outlines by which they are bounded.

[1] Reference will be afterwards made to one or two interesting exceptions to this rule, the Caspian Sea being the chief.

8. The surface of the sea forms a sharp, well-marked platform, which is called the **sea-level**. It serves as the line from which the heights of the land and the depths of the oceans are measured. Slight differences of level have been detected between different oceans or seas, as, for instance, between the Atlantic and Pacific, on the two sides of the narrow part of America, and between the Mediterranean and the Red Sea. But such differences do not amount to more than a few feet, so that for most practical purposes we may assume the sea-level to be uniform.

9. Passing next to the land, we observe from the map that while massed on the whole in the northern half of the globe, it is far from forming there one solid block; on the contrary, it is intersected by branches from the sea so as to be easily grouped into a few great subdivisions. These are termed **Continents**. Strictly speaking, there are only two continents, the Old World and the New World (Art. 5). More usually, however, they are grouped in three pairs, the first pair consisting of North and South America, the second of Europe and Africa, and the third of Asia and Australia.

10. Now, in considering the general arrangement of the land on the globe, we soon see that one of the most marked general features of the continents is their tendency to be massed together towards the north, and to taper away towards the south to about the parallel of 45° S. Look, for example, at the way in which the mass of Africa dwindles away southward to the Cape of Good Hope, and how the huge bulk of the Asiatic Continent and its prolongation in Australia tapers away to the headlands of Tasmania. Still more remarkable is the attenuation of the American Continent and its sharpened termination at Cape Horn.

11. Again, not only is the land much intersected by inlets, such as the great Mediterranean Sea and the Red Sea, but unlike the sea (Art. 5) large parts are completely cut off from the main mass, as in Australia, New Zealand, Japan, Britain, and the host of **islands**, large and small. Other parts nearly isolated are termed **peninsulas**. Of these Africa may be taken as the most conspicuous illustration. It is joined only by the little connecting neck or **isthmus** of Suez to the continent of Asia, so that were that strip of lowland to be cut through or sunk beneath the sea, Africa would become an island. This, indeed, has been in some sense accomplished by man in the making of the Suez Canal.

12. Another feature of the land marks it off in striking contrast from the sea. The surface of the latter, though liable to be roughened by ripples and waves, preserves everywhere the character of one vast plain. But the surface of the land abounds in unevenness. Some parts, indeed, are flat, but most of it undulates into hills and valleys, and some portions mount up into vast ranges of rugged and pointed mountains.

13. This irregular distribution and inequality of surface give the land a peculiar character and influence in the physical geography of the earth's surface. Without inquiring at present how this influence shows itself, we can readily see that if either in the air or in the sea there is any movement of circulation, the currents, both of air and sea, must be greatly affected by the position and form of the continents and islands which may lie in their course. A current in the sea moving westward, for instance, across the middle of the Atlantic, will strike against the long ridge of America, and be turned aside either to right or left, or both. A current of air, on the other hand, if it should take its rise from some centre in the ocean region, will be turned aside when it meets a mass of high land, and may be driven to ascend the slopes, discharge its moisture, and flow on at a much greater height, and at a very different temperature (Lesson X. Art. 31). The varying influence of sea and

land upon the air profoundly affects the different climates of the earth.

14. It has been already stated (Lesson III. Art. 4) that at many parts of our planet's surface, pipes or funnels exist, which descending deep into the earth, cast out from time to time steam, hot vapours, dust, stones, and melted rock, so as in the end to form conical hills or mountains called **volcanoes**. It is observed that orifices of this kind are apt to occur in long lines, and especially along the back-bones of the continents, and in long chains of islands (see Plate IX.). Over many parts of the globe, too, and particularly in those regions where volcanoes abound, the ground is frequently shaken by **earthquakes**, and sometimes permanently lifted above its previous level. Such appearances as these afford indications, not only of the nature of the earth's interior, but also of the way in which the interior affects the surface. They help us to understand how it is that the land has been ridged up above the general level of the sea.

15. Into these matters we shall enter more fully in later Lessons. Meanwhile, carrying with us this general outline of the several parts of the earth, we may proceed to consider these parts one by one in detail.

CHAPTER II.

AIR.

Lesson VI.—*Its Composition.*

1. Above and around us, to what part soever of the earth's surface we may go, at the top of the highest mountain, as well as at the bottom of the deepest mine, we find ourselves surrounded by the invisible ocean of gas and vapour which we call AIR. It therefore wraps the whole planet round as an outer envelope. Considered in this light it receives the distinctive name of the ATMOSPHERE, that is, the vapour-sphere—the region of clouds, rain, snow, hail, lightning, breezes, and tempests. In the study of the earth as a great habitable globe this outer encircling ocean of air is the first thing to be considered. What is it? and what purposes does it serve in the general plan of the earth?

2. In early times men regarded the air as one of the four elements out of which the world had been made. It is not so very long since this old notion disappeared. But now it is well known that the air is not an element, but a mixture of two elements—viz. the gases called **Nitrogen** and **Oxygen**. It is easy to prove this by burning a piece of another element, phosphorus, in a closed jar. We thereby remove the oxygen, which unites with the phosphorus to form a compound substance, and the nitrogen is left behind.[1] In various ways chemists have analysed or decomposed air into its component

---

[1] See Roscoe's *Chemistry Primer*, p. 21.

elements, but the result is always the same, viz., that in every hundred parts of ordinary air there are by weight about seventy-nine of nitrogen and twenty-one of oxygen.

3. Air, when carefully tested, is always found to contain something else than nitrogen and oxygen. Solid particles, with various gases and vapours, are invariably present, but always in exceedingly minute, though most irregular quantities, when compared with the wonderfully constant proportions of the two chief gases. Some of these additional components of air are not less important than the nitrogen and oxygen. That they exist may be easily proved, and some light may thereby be thrown on the nature and uses of the air.

4. The presence of vast numbers of **solid particles** in the air may be shown by letting a beam of sunlight, or of any strong artificial light, fall through a hole or chink into a dark room. Thousands of minute motes are then seen driving to and fro across the beam as the movements of the air carry them hither and thither. Such particles are always present in the air, though usually too small to be seen, unless when, as in the darkened room, they are made visible against surrounding darkness by the light which they reflect from their surfaces when they cross the path of any strong light-rays. They are quite as abundant in the dark parts of the room, though for want of light falling upon them they are not seen there.

5. Could we intercept these dancing motes and examine them with a strong microscope, we should find them to consist chiefly of little specks of dust. But among them there sometimes occur also minute living germs, from which, when they find a fitting resting-place, lowly forms of plants or animals spring. Various diseases arise from such infinitesimal germs, which are so small as to pass with the air into our lungs, and thus to reach our blood, where they rapidly multiply.

6. It is difficult to catch these tiny motes from a sunbeam, but rain does this admirably for us. One great office of rain is to wash the air and free it from impurities. Hence, when rain-water is carefully collected, especially in large towns, it is found to contain plenty of solid particles, which it has brought down with it in its fall through the air. The accompanying drawing, for example, shows what is seen when a small quantity of rain, gathered from an open space in a town, is evaporated to dryness, and the residue is placed under a microscope. Abundant particles of dust or soot may be observed, mingled with minute crystals of such substances as sulphate of soda and common salt. Hence we learn that, besides the solid particles, there must be floating in the air the vapours or minute particles of various soluble substances which are caught by the rain and carried down with it to the soil. In seizing these impurities and taking them with it to the ground, the rain purifies the air and makes it more healthy, while at the same time it supplies the soil with substances useful to plants.

FIG. 7.—What is seen after some raindrops collected in a town are evaporated, and the residue is placed below a microscope.

7. But far more important than these solid ingredients are three invisible substances,—ozone, carbonic acid gas (carbon dioxide), and water-vapour. After a thunderstorm the air may sometimes be perceived to have a peculiar smell, like that which is given off from an electric machine. This is **ozone**, which is believed to be oxygen gas in a peculiar and very active condition. It promotes the rapid decomposition of decaying animal or vegetable matter, uniting with the noxious gases, and thus disinfecting and purifying the air. It is most abundant where sea-breezes blow, and least in the air of the crowded parts of towns. The healthiness or unhealthiness of the air seems to depend much on the

quantity of ozone, which may be estimated by the amount of discolouration produced by the air within a certain time upon a piece of paper prepared with starch and iodide of potassium.

8. When a piece of coal is set on fire it burns away until nothing but a little ash is left behind. Or when a candle is lighted it continues to burn until the whole is consumed. Now, what has become of the original substance of the coal and the candle? It seems to have been completely lost; yet in truth we have not destroyed one atom of it. We have simply, by burning, changed it into another and invisible form, but its component elements are just as really existent as ever. We cannot put these back into the form in which they appeared in the coal and candle, but we can at least show that they are present in the air, where they pass mainly into water-vapour and other gaseous compounds, of which the most important is **carbonic acid gas** or **carbon dioxide**.

9. The substance of a piece of coal or of a candle is composed of different elements, one of which is called *carbon*. This element forms one of the main ingredients out of which the substance of all plants and animals is built up. Our own bodies, for example, are in great part made of it. In burning a bit of coal, therefore (which is made of ancient vegetation compressed and altered into stone), or a candle (which is prepared from animal fat), we set free its carbon, which goes off at once to mix with the air. Some of it escapes in the form of little solid particles of soot, as we may show by holding a plate over the candle flame, when the faint column of dark smoke at once begins to deposit these minute flakes of carbon as a black coating of soot on the cool plate. The black smoke issuing from chimneys is another similar illustration of the way in which solid particles are conveyed into the air.

10. But the largest part of the carbon does not go off in smoke. In the act of burning it is seized by the

oxygen of the air, with which it enters into chemical combination, to form the invisible carbonic acid gas. It is, indeed, this very chemical union which constitutes what is called burning or combustion. The moment we prevent the flame from getting access of air, it drops down and soon goes out, because the supply of oxygen is cut off.[1] All ordinary burning substances, therefore, furnish carbonic acid gas to the atmosphere.

11. The amount of this gas which the atmosphere thus obtains is, of course, comparatively small, for the quantity of vegetable or animal substance, burned either by man or naturally, must be but insignificant when the whole mass of the atmosphere is considered. A vastly larger quantity is furnished by living air-breathing animals. In breathing we take air into our lungs, where it reaches our blood. A kind of burning goes on there, for the oxygen of the air unites with the carbon of the blood; carbonic acid is produced, and comes away with the exhausted air, which we exhale again before taking the next breath. Just as a burning candle is extinguished when a glass is so placed over it as to exclude the air, so a living animal is killed when it is shut into a confined space from which the access of the outer air is cut off. When we reflect that every air-breathing animal is continually supplying carbonic acid gas to the atmosphere, we perceive how important this source of supply must be.

12. Living plants in the presence of sunlight have the power of abstracting from the carbonic acid of the air the carbon of which their framework is so largely made.[2] When they die their decay once more sets loose the carbon, which, uniting again with oxygen, eventually

---

[1] For some simple experiments on the nature and production of carbonic acid, see Roscoe's *Chemistry Primer.*

[2] The familiar practice of growing cress-seed on moist cloth shows this power of living plants. Kept in the light the seeds soon begin to grow, and yield a crop of cress, the carbon of which the grown plants consist being derived, not from the water, nor from the cloth, but from the air. (Roscoe's *Chemistry Primer.*)

becomes carbonic acid gas, and is carried by rain into the soil, or is taken up by the air. All decaying plants and animals, freely exposed to air and moisture, furnish this gas.

13. Lastly, in many parts of the world, particularly in volcanic regions, this same gas is given out in large quantities from the ground. From all these various sources, then, the atmosphere is continually replenished with carbonic acid gas to supply the loss caused by the enormous demands of the vegetable world for carbon.

14. Nevertheless, the quantity of this gas present in the air is very small compared with the volume of the nitrogen and oxygen. It has been found to amount to no more in ordinary pure air than about three parts in every ten thousand of air. Yet this small proportion suffices to support all the luxuriant growing vegetation of the earth's surface.

15. By the term **water-vapour**, or **aqueous-vapour**, is meant the invisible steam always present in the air. Every one is familiar with the fact, that when water is heated it passes into vapour, which becomes invisibly dissolved in the air. A vessel of water, for instance, may be placed on a table in the middle of a room, heated by means of a spirit lamp till it boils, and kept boiling till the water is entirely driven off into vapour or **evaporated**. The air in the room shows no visible change, though it has had all this water-vapour added to it. But it may be easily made to yield back some of the vapour. Let an ice-cold piece of glass, metal, or any other substance be brought into the room. Though perfectly dry before, its surface instantly grows dim and damp. And if it is large and thick enough to require some minutes to get as warm as the air in the room, the dimness or mist on its surface will pass into trickling drops of water. The air is chilled by the cold glass, and gives up some of its moisture. Cold air cannot retain so much diffused vapour as warm air, so that the capacity of the

air for vapour is regulated by its temperature. (See Lesson X.)

16. It is not needful to boil water in order to get water-vapour enough in the air of a room to be capable of being caught and shown in this way. In a warm sitting-room, where a few persons are assembled, there is always vapour enough to be made visible on a cold glass. In frosty weather the windows may be found streaming with water inside, which has been taken out of the air by the ice-cold window-panes. Whence came this moisture? It has been for the most part breathed out into the air by the people in the room.

17. Each of us is every moment breathing out water-vapour into the air. As a rule, we do not see it, because the air around us is dry enough to dissolve it at once. But anything which chills our breath will make the vapour visible, such as breathing on a cold piece of glass, or metal, when a film of mist at once appears on the object; or walking outside on a cold or very damp day, when the vapour of each breath becomes visible as a little cloud of mist in the air.

18. No matter how dry the air may appear to be, more or less of this invisible water-vapour is always diffused through it. Every mist or cloud which gathers in the sky—every shower of rain, snow, or hail, which falls to the ground—every little drop of dew which at nightfall gathers upon the grass, bear witness to its presence.

19. The importance of this ingredient of the atmosphere in the general plan of our world, can hardly be over-estimated. It is to the vapour of the atmosphere that we owe all the water-circulation of the land—rain, springs, brooks, rivers, lakes—on which the very life of plants and animals depends, and without which, as far as we know, the land would become as barren, silent, and lifeless as the surface of the moon. It is, likewise, to the changes in the supply of this same invisible, but ever

present substance, that the rise of winds and storms is largely due (Lesson XI.).

20. The quantity of water-vapour in the air, always comparatively small in amount, constantly varies from day to day, and from hour to hour. It is affected by each change of temperature, and by every condensation and evaporation that takes place. The air is never quite free from vapour, even in the driest climates. The proportion of vapour increases with temperature, as explained in Lesson VIII.

LESSON VII.—*The Height of the Air.*

1. Though no actual upper limit has been found to the atmosphere corresponding at all to the sharply-defined surface of the sea, we cannot suppose the air to extend indefinitely outwards from the earth. The atmospheric envelope clasps the planet firmly, and moves along with it, both in rotation and revolution. Were this not the case, it is plain that the earth's movement through the air would be far more rapid than the most furious hurricane. No loose object could appear on the surface of the globe without being instantly whisked off. But the attraction of the earth retains the atmosphere in its place, so that it is carried with the rest of the planet through space.

2. There must then be some upper limit to the atmosphere. Beyond it lies the ether, which is supposed to fill all space, and through which all the heavenly bodies, and the rays of light proceeding from them are moving. How can it be known how far the atmosphere extends above us?

3. There are different ways of trying to answer this question. Let us look for a moment at one of them. On clear, dark nights, shooting-stars or meteors may be seen, sometimes in considerable numbers. They suddenly appear, and after making a train or tail of light, quickly vanish. Sometimes they have been actually heard to explode in the sky, and fragments of them have been picked up. They have been carefully watched by astronomers. By observing their positions and directions of movement from different stations, it has been possible to determine how high they are above us, by a process very similar to that by which distances are estimated on the earth by taking angles to any object from the two ends of a measured base-line. They have thus been found to begin to be visible at from 70 to 500 miles from the earth's surface.

4. In themselves they are little fragments, usually not more than a few ounces or a few pounds in weight; but they revolve round the sun with the velocity of planets or of comets. While still following their usual orbit, they are in themselves cold,[1] opaque bodies. The light which they emit, and by which they become visible, arises from the fact that, drawn out of their course by the attraction of our earth, and rushing into our atmosphere with an enormous velocity, they rapidly get heated by friction against the air, as well as by the heat liberated from the compression of the air in front of them. They soon become white-hot, and in most cases so intense is the heat to which they are raised that they are dissipated into vapour, which shows as a tail or train of light, gradually fading out of the sky. From the height at which these shooting-stars begin to glow, it is inferred that the atmosphere must extend at least to 500 miles above the general solid surface of the earth, and may even reach to 600 miles.

5. But the air at these great heights must be very different in many respects from the air next the earth. We could not breathe it. When, for instance, travellers ascend lofty mountains, they often find an increasing

[1] They probably have nearly the temperature of space, which is supposed to be about 271° Fahr. below the freezing-point of water.

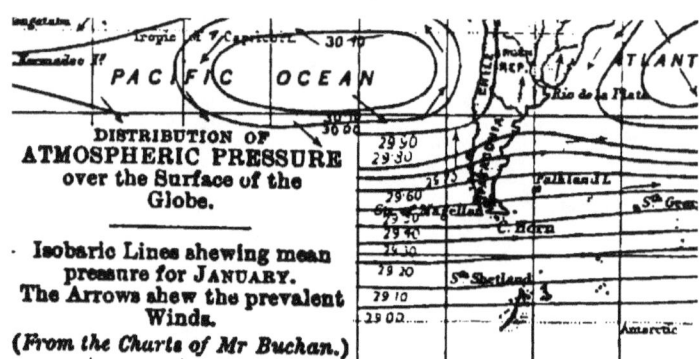

DISTRIBUTION OF
ATMOSPHERIC PRESSURE
over the Surface of the
Globe.

Isobaric Lines shewing mean
pressure for JANUARY.
The Arrows shew the prevalent
Winds.
(*From the Charts of Mr Buchan.*)

difficulty in breathing as they advance. In a similar manner persons who have gone up to great heights in balloons have become insensible, and have nearly died from the difference between the air below and that above. The chief difference is in density, the air getting continually lighter or thinner as it recedes from the sea-level. Our bodies cannot endure the difference between the close, heavy air to which we are accustomed next the earth, and the thinner air farther up—though when the transition is made gradually, the lungs may be trained to breathe the air at considerable heights on lofty mountains. Breathing, however, becomes impossible at a greater height than six or seven miles. Beyond that distance the air must get less and less dense, until it reaches the extremest attenuation at the farthest outskirts of the atmosphere.

LESSON VIII.—*The Pressure of the Air.*

1. Though invisible to us, and so light that we live and move in it without thinking of its existence, the atmosphere nevertheless presses upon every part of the earth. The lower layers must be pressed down by the weight of all the mass of air above them. This is what is usually spoken of as **atmospheric pressure**, which is due, not merely to the weight, but to other properties of the gases and vapours which form the air.

2. To prove the pressure of the air, take a little glass phial, and putting it to the mouth, suck out the air as well as possible, taking care to let the tongue press instantly back upon the opening. You feel the tongue driven into the phial, and perhaps even with pain, owing to the pressure of the air outside, and the absence of corresponding pressure from air inside. An effect of this kind is capable of being accurately measured. Accordingly, observation shows that, at the sea-level,

this pressure is as much as about 14¾ pounds upon every square inch. Every one of us, therefore, bears a weight of 12 or 14 tons of air. Yet we do not feel this pressure, because it is exerted equally on all sides, and because the air within our bodies has the same pressure outwards which the air outside has inwards. If we could withdraw the air from all the cavities and passages in a human body, the weight of the air outside would crush the body in and cause immediate death.

3. Bearing in mind that each part of the atmosphere has to bear the weight of all the air which lies above it, we can understand why, as we ascend a mountain, and have a decreasing mass of air above us (Lesson VII. Art. 5), we should encounter air of less and less density. Atmospheric pressure diminishes with altitude.

4. The pressure of the atmosphere is measured with the instrument called the **Barometer**. The principle of this instrument is that the weight of the atmosphere will balance the weight of a column of any other fluid, the height of that column being determined by the relative weight, or what is called the specific gravity, of the fluid employed. A glass tube about thirty-three inches long, closed at one end, is filled with mercury, and then inverted with its open end in a cup of the same fluid metal. The mercury is observed to fall in the tube until, if near the sea-level, it stands at a height of about thirty inches above the surface of the mercury in the cup. The column of mercury, thirty inches in height, is balanced and kept from descending farther by the pressure of the overlying column of the atmosphere upon the surface of the mercury in the cup. The more the atmosphere presses upon the latter, the farther does the mercury rise in the tube, while on the other hand the less it presses, the more the mercury sinks in the tube.

5. With such an instrument, variations in atmospheric pressure may be detected, even when so small and slow that we should never otherwise have been sensible of

them. If the height of the mercury in the tube is accurately noted at starting, and the barometer is carried to a higher level, the mercury will be observed to fall because of the diminished pressure, while it will rise again when the instrument is brought back to lower ground. So regular and delicate is this action that the barometer is often employed for measuring heights. Were there no other cause for variations in the pressure of the atmosphere than merely elevation, the barometer would be most useful for that purpose, but, as we shall immediately see, would be of no service in the way it is now chiefly employed.

6. The diminution of atmospheric pressure according to height in the air is regular and constant. But besides this, the pressure is liable to continual changes at every level, sometimes sudden and great, at other times slow and slight. We are made sensible of these variations when they are accompanied by changes of weather. But they are most accurately measured by the movements of the barometer. If from any cause the pressure should decrease, the mercury will fall; should it increase, the mercury will rise; the rapidity or slowness of the movement in the column of mercury affording us, as it were, a mirror of the amount and the rate of change in the balancing atmospheric column.

7. Suppose, by way of illustration, that on looking at the barometer some morning, we should find the mercury to have fallen a whole inch during the night; the mercury column would thus indicate that during a few hours it had lost a thirtieth part of its whole length, and we should be justified in believing that, in some way or other, the column of air that presses on the mercury in the cup, had in like manner lost a thirtieth part of its pressure or weight. Some of the upper portions of the atmosphere must have flowed over into surrounding regions so as to diminish the pressure to this extent. No such sudden and great change, however, could fail to

produce a violent hurricane. The fall of the barometer in almost every case occurs in time to prepare us for the coming storm.

8. The barometer tube is divided into inches, and these into tenths, hundredths, and thousandths, so that the position of the mercury is noted to the thousandth of an inch. When the pressure of the atmosphere just balances a column of mercury thirty inches in height, the barometer is said to mark or stand at 30·00. If the mercury falls half an inch, the barometer stands at 29·50. If it then rises a tenth of an inch, it is read as 29·60, a further rise of a hundredth of an inch would make 29·61. The height of the mercury in the barometer at the level of the sea has been found to be very nearly thirty inches as a mean over the whole globe. In different regions, however, the actual average height for the year varies considerably from that mean. In the Pacific Ocean, for instance, some distance to the westward of California, the mercury stands on the average at 30·30 inches. On the other hand, in the north-west of Iceland it stands at a mean height of 29·66 inches, while within the Antarctic circle the average is even lower. When the mercury has fallen below its average height, it indicates a **low pressure**, when it rises above the average it marks a **high pressure**.

9. The importance of noting carefully the variations in atmospheric pressure will be apparent from the fact, which has now been proved by observation in all parts of the world—that it is differences of pressure which give rise to winds, storms, and in short all the movements of the air which are intimately connected with changes of weather.

But what is the cause of these variations in pressure? Why should the air be liable to such increasing and often sudden, as well as great changes? The answer to these questions is, that the pressure is affected, 1st, by temperature, and 2d, by aqueous vapour.

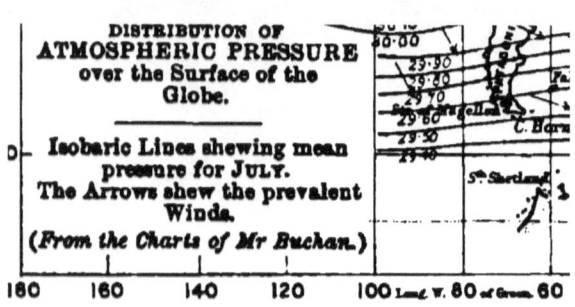

DISTRIBUTION OF ATMOSPHERIC PRESSURE over the Surface of the Globe.

Isobaric Lines shewing mean pressure for JULY.
The Arrows shew the prevalent Winds.
(*From the Charts of Mr Buchan.*)

PLATE III.

10. (1) **Temperature.** It is not difficult to see how this influence acts. When air is heated it expands, when cooled it contracts, behaving in this respect like other substances. Cold air is denser than warm air, so that the latter ascends, while the former descends. The ascent of warm air must necessarily diminish atmospheric pressure. When a broad tract of the earth's surface, such for instance as the centre of Asia, is greatly heated by the sun's rays, the hot air in contact with the ground rises and flows over into the surrounding regions. Hence the atmospheric pressure there is lowered during the hot months of the year.

11. (2) **Aqueous Vapour** is, however, found to be a still more important agent in affecting the pressure of the air. We have already considered how universally present this invisible vapour is, and how easily it may usually be made visible by cooling the air; when it is at once changed into visible water. In Lesson X. an account is given of the vapour of the atmosphere; how it is continually passing into the air, and as constantly passing out of it again into some visible form of water. Let us consider here how this unceasing process affects the pressure of the atmosphere.

12. Suppose that we take two empty glass vessels, each capable of containing exactly one cubic foot of any substance. By means of an air-pump we remove the air from them as completely as possible; one of them we fill with water-vapour, at a temperature of, say 50° Fahr. (Lesson IX. Art. 1); the other, at precisely the same temperature, we fill with perfectly dry air, that is, air from which the water-vapour has been removed as thoroughly as possible. We then weigh them both, and after allowing for the weight of the glass-vessel in each case, we find that the vapour weighs only 4·10 grains, while the air weighs as much as 546·81 grains.

13. Now, without entering into the question how far atmospheric pressure is due to mere weight or to other causes, what conclusion must we draw from this experiment? Evidently, that water-vapour is vastly lighter, or has less pressure, than air. At the temperature of 50° it is about 133 times lighter, and though the difference would be rather less at higher temperatures, still, under all the temperatures to which the atmosphere is ordinarily subject, the weight or pressure of the air is always far greater than that of the vapour.

14. Suppose, again, that we take six glass vessels, each holding exactly a cubic foot, and fill them as follows:—three with air saturated with vapour, each at a different temperature, the first, say at that of freezing-water (32°), the second at that of an ordinary English spring morning (50°), and the third at that of a warm English summer noon (80°), and three with perfectly dry air at the same three temperatures. In the first three, the air of each vessel contains as much vapour as its temperature will allow it to retain, and as we have seen that cold air cannot hold so much as warm air, we know that in the warmest vessel there must be a good deal more vapour than in the coldest. We weigh them all carefully as before, and find that the ice-cold moist air weighs about a grain and a quarter (1·27 grain) less than the perfectly dry air at the same temperature, that the moist air at the medium temperature is about two and a half grains lighter than the dry air, and that the warmest moist air is about six and a half grains lighter than the warmest dry air.

15. What lesson does this experiment teach? Plainly that the addition of water-vapour makes air lighter or diminishes its pressure, and that this change is greater the warmer the air, for more vapour can be dissolved in warm than in cold air.

The vapour which rises so abundantly from sea and land into the atmosphere diffuses itself through the air, pushing the air atoms aside; and having far less weight than the air, it necessarily lessens the density of the atmosphere, or in other words, lowers the atmospheric pressure. A mixture of air and vapour is lighter than the same volume of dry air would be, and, of course, the larger the proportion of vapour, the greater will this difference be.

16. The quantity of vapour in the atmosphere is constantly varying from day to day, and from season to season (Lesson X.). We perceive, therefore, that this cannot fail to be at least one grand cause of the ceaseless oscillations of pressure which the barometer reveals. The addition of a large volume of vapour to the atmosphere lowers the atmospheric pressure, and the mercury in the barometer therefore falls. The removal of this vapour either by condensation into rain, or otherwise, restores the pressure, and the mercury rises again. Sometimes these fluctuations are very slow, extending over days or weeks, sometimes a great oscillation may take place within a few hours.

17. How it is that these vast accumulations of vapour in any part of the atmosphere are brought about is still unknown. It is well ascertained, however, that with the condensations that follow, they determine movements of the air. When sudden and extensive, they are accompanied by storms of rain and wind. When less violent, they still show their influence upon the winds and weather.

18. All movements of the atmosphere arise from differences of pressure, caused, as far as we can tell, by variations in temperature, and water-vapour. We shall consider these in the next two lessons.

19. Observations with the barometer carried on for many years in all quarters of the globe, have enabled meteorologists to prepare charts showing, by means of lines of equal pressure called **Isobars**, the general distribution of atmospheric pressure over the earth's surface for each month or season, or for the whole year. Plates II. and III. are examples of such charts. It is found that, taking a very broad view of the subject, there are three great areas of low pressure. One extending as a broad belt round the equatorial regions, the other two lying about each pole; and two areas of high pressure extending parallel to the equatorial belt on either side, and separating it from the polar area of low barometer. These areas are most continuous in the southern hemisphere; but even there, and therefore still more in the northern, they are broken up into detached portions owing to the irregular distribution of sea and land. They change position too with the seasons, as is well shown by comparing the distribution of pressure in January with that in July, as is done in Plates II. and III.

LESSON IX.—*The Temperature of the Air.*

1. As in the study of the pressure of the air so in that of its temperature, it would not be possible to make much progress without some means of accurately measuring the fluctuations, for it is only the more marked of these of which we are sensible from the comfort or discomfort they bring to us. Happily, in this case too, the instrument for accurate measurement is exceedingly simple both in its construction and use. It is called the **Thermometer**, or heat-measure, and consists of a small glass tube closed at both ends, the lower of which is expanded into a bulb. The tube, previously deprived as completely as possible of air, is partially filled with mercury or spirits of wine, and attached to a flat piece of ivory, wood, or other substance, on which is placed a graduated scale. Under the influence of heat the fluid in the tube expands and rises; when the heat is withdrawn it contracts and descends. The temperature is expressed by the figure on the scale of degrees, opposite to which the upper end of the mercury column stands. The thermometer scale in most common use in this

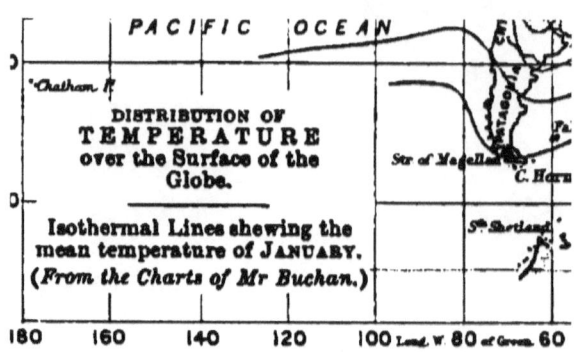

DISTRIBUTION OF
TEMPERATURE
over the Surface of the Globe.

Isothermal Lines shewing the mean temperature of JANUARY.
(*From the Charts of Mr Buchan.*)

PLATE IV.

country, called after the original maker, Fahrenheit's, is so graduated that when the instrument is placed in melting ice or in water just about to freeze, the mercury marks 32°.[1] This is the *freezing-point* of fresh water. On a pleasant summer day in England the mercury may mark 70°. On a hot noon in India it would rise to 90° or more, while on the burning sands of an African desert, at the hottest time of the day, it might sometimes stand as high as 150°, or even higher. Under ordinary atmospheric pressure the thermometer indicates 212° as the temperature at which water boils. When the mercury is low in the tube, it marks cold, or what is called a **low temperature**; when it stands high in the tube it indicates heat, or a **high temperature**.

2. By means of the thermometer it is possible to measure very minute changes of temperature, and to compare the range of temperature in different places. Observations of this kind have now been carried on for many years in all parts of the world, with the result of making known the general distribution of temperature over the globe. To show this distribution it is usual to construct maps on which lines are drawn through all places having the same temperature (see Plates IV. and V.). Such lines have received the name of **Isotherms**, **Isothermal** lines, or lines of equal temperature. Each of them is named after the degree of the thermometer which it expresses, as, for example, the isotherm of 60°, which shows that all the places through which it is drawn on the map have the average temperature of 60°.

3. Whence does the earth receive its heat, and why does the temperature of one part of its surface vary so much from that of another?

[1] The zero (0°) of Fahrenheit's scale is 32° below the freezing-point of fresh water. When the temperature sinks below zero it is marked by prefixing the *minus* sign, thus -5°, -10°: that is five, ten degrees below the zero-point.

4. Although, as we have seen (Lesson III.), our planet was once probably a molten globe, and retains even now a vast amount of heat in its interior, its surface temperature is not materially affected thereby. Were it left without any other source of heat than its own, its surface would become so intensely cold as to be utterly uninhabitable by at least the races of plants and animals which now live upon it.

5. It is from the Sun that our supply of heat comes; and, according to the greater or less amount of this heat which different parts of the earth receive, they vary in temperature. Were there no water-vapour in the atmosphere the heat-rays would pass through with no appreciable diminution; but the presence of this vapour partially arrests them, and thereby increases the temperature of the air, but only to a very limited extent. Hence in dry climates radiation (Art. 6) is most active, the days being warmer and the nights colder than in moist climates.

6. There are three processes whereby temperature is interchanged. (1) The surface of the land, being warmed by the sun's heat, warms the layer of air that lies upon it. This is called **conduction**. (2) The warmed air rises, while colder air descends, to be in turn warmed and made to ascend. Currents in the atmosphere are thus produced, and the heat carried along by them (or by currents in the sea) is said to be diffused by **convection**. (3) Heat is continually passing to and fro between objects freely exposed to each other. This is known as **radiation**. The sun, for example, is unceasingly radiating heat in all directions, and so likewise is our own globe. During the day the illuminated side of the earth receives much more heat from the sun than it gives off itself, and is consequently warmed. At night, on the other hand, the dark side of the earth radiates into space more heat than it receives from planets and stars, and is therefore cooled. It is when the sun's rays have heated a

part of the earth's surface that the air resting upon that surface becomes heated by conduction, and the currents arise which diffuse the heat by convection.

7. The heating power of the sun's rays has been ascertained to be dependent upon the angle at which they reach the surface of our planet. Wherever they fall vertically upon the earth, as at B in Fig. 8, their heating power is greatest. This power diminishes as the direction of the rays recedes more and more from the vertical, until, when they become horizontal, as at A and C, it reaches its lowest point. Hence, though powerful at

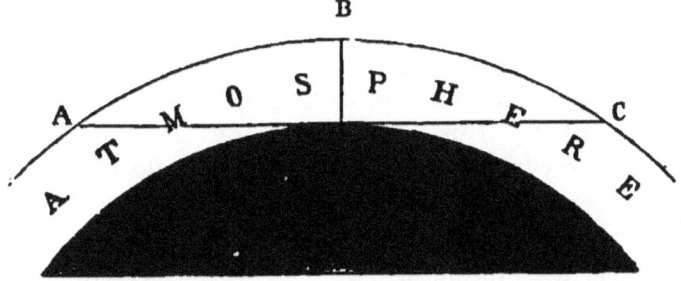

Fig. 8.—Diagram showing the influence of the varying thickness of the atmosphere in retarding the Sun's heat. A. Line of the Sun's rays at sunrise. B. Line of the rays at noon. C. Line of the rays at sunset.

noon, they are comparatively feeble in the morning and in the evening.

8. In considering, then, the manner in which the temperature of the atmosphere is distributed over the globe, we see at the outset that those countries must necessarily be warmest where the solar rays are vertical or most nearly so, and that those regions must be coldest where the rays are most oblique. At each place between the tropics (Lesson II. Art. 11) the sun is vertical twice a year. That belt of the globe must needs therefore have the highest temperature. Round the poles the sun does not shine for six months in winter, and never gets very high in the sky even in summer. These must consequently be the coldest regions. Hence the

first conclusion we may draw regarding the temperature of the atmosphere is that it depends upon distance from the equator, or, expressed more shortly (1) **temperature is regulated by latitude.**

9. If no counteracting influence came into play there would be a regular diminution of temperature from the equator to the poles. Every latitude would then have its own temperature, so that a determination of the average temperature of a place would give us the latitude; or, knowing the latitude, we should know also what must be its temperature.

10. But this correspondence holds only to a very limited degree. The places which may chance to lie on the same parallel most frequently do not enjoy the same temperature. London, for example, coincides in latitude with Labrador. But in England the summers are not hot, and the winters are not very cold, while in Labrador the summers are mild and the winters rigorously cold. There must, therefore, be some other cause at work modifying the influence of latitude. What is this secondary cause?

11. When Isothermal lines are drawn round the globe, so as to connect together all the places which have the same mean temperature, these lines, instead of following the parallels of latitude, are found to bend up and down, and these bendings are observed to have reference to the positions of the continents and oceans. Any temperature may be chosen. We may select, for example, a temperature of 32° Fahr. or of 40°, 50°, 60°, or any other arbitrary point on the thermometer, it being understood that over every part through which the line is drawn the mean temperature during the year or one of the seasons or months has been found by prolonged observation to be the same, and amounts to the number of degrees fixed upon. In this way we may represent the mean temperature of the whole year, or that of winter or of summer.

12. By referring to the maps in Plates IV. and V., which represent the distribution of temperature over the globe for January and for July, it will be seen how these lines are drawn and how simply they tell the story of the arrangement of temperature upon the earth. They are least irregular in the southern hemisphere over the wide expanse of ocean; they show the greatest deflections across North America, the Atlantic, Europe, and Asia. In this way they indicate that temperature is more uniform and more directly dependent on latitude in the oceanic parts of the globe than in the continental, or in the regions where the oceanic and continental come together, as they do in the basin of the Atlantic.

13. In order to see the full use and meaning of these isothermal lines, let us take by way of illustration the line in the northern hemisphere marking a mean annual temperature of 50° Fahr. Traced over Britain, this line runs from about London, across the centre of England and the north of Wales: that is to say, all the parts of the country lying along that line have a mean annual temperature of 50°, while the districts to the north-east are a little colder and those to the south-west a little warmer. The line bends south-westward and crosses to the western shores of Ireland. If now we turn to the opposite side of the Atlantic to see where places are to be found having the same average temperature for the year, we do not meet with them on the same parallel of latitude as in the British Islands. They lie much farther south, so that the line or isotherm of 50° makes a bend in crossing the ocean, and reaches the American coast not far from New York. The mean annual temperature of London and New York is the same. And yet New York is about as far south from London as Madrid.

14. This isothermal line of 50° between Europe and America, and still more the other lines lying to the north of it, illustrate how far the belts of equal heat are from coinciding with the parallels of latitude. It will be seen from the maps how this divergence is determined by the way in which the areas of land and sea are grouped.

15. Land is sooner heated by the sun's rays than the sea, and also gives off its heat again sooner. The sea, though it does not become so hot as the land does, retains its heat longer, and is enabled, by virtue of its liquidity and motion, to diffuse it. Hence, the influence of the sea tends to mitigate both the heat and the cold of the land. Its warm currents heat the air resting on them, and so give rise to warm winds which blow upon the land, while its colder waters in like manner temper the air, which reaches the land in cooling breezes, or it may be in cold damp winds and fogs. Thus, in the basin of the North Atlantic, a warm ocean-current called the Gulf Stream issues from the Gulf of Mexico, and, augmented by the surface-drift of warm water which is driven onward by the prevalent south-west winds, flows across the Atlantic to the shores of Britain and even of Spitzbergen. It brings with it the supplies of heat which make the climate of the west of Europe so much less cold than it would naturally be. On the other hand, an icy stream of water, coming out of Davis Strait, bears a chill to the coasts of Labrador and Newfoundland. The ocean, therefore, by its cold currents is depressing the temperature in America, along the same latitudes where, in Europe, by its warm currents, it is raising it.

16. A large body of land in high latitudes, by accumulating snow and ice, gives rise to a lower temperature, and a similar mass of land in low latitudes, by exposing a broad surface to the tropical rays of the sun, produces a higher temperature than would be found if the region were occupied by the sea. In illustration of this statement it will be seen from the map (Plate IV.) that the January isothermal lines marking temperatures below the freezing-point, come a good way southwards over Northern Asia, and again over Greenland and North

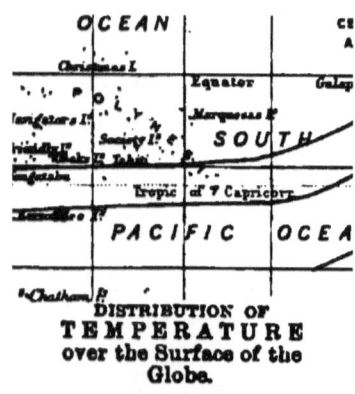

DISTRIBUTION OF
TEMPERATURE
over the Surface of the
Globe.

Isothermal Lines shewing the
mean temperature of JULY.
(*From the Charts of Mr Buchan.*)

PLATE. V.

America, though in the water passage between these regions they stretch a long way northwards. The coldest winter temperature on the globe is that of north-eastern Siberia, where the thermometer falls to $-55°\cdot 8$ (nearly 56° below zero or 88° below the freezing-point). The map of the July temperature (Plate V.) shows that over the equatorial parts of America and the great mass of Africa and Southern Asia, the space enclosed between the isothermal lines of 80° swells out greatly, so as to include a much larger area than where these lines cross the oceans. The hottest climates are found in the dry regions that stretch from the African deserts through Arabia to India.

17. It by no means follows that two places which have the same average yearly temperature enjoy the same climate. For instance, Reykjavik, in the south of Iceland (Lat. 64° 40′ N.) has a mean annual temperature of about 40° Fahr., while the city of Quebec has one of about 39°; but in July the average temperature in the former locality is 52°, in the latter it is 67°; while in the one case the January temperature is 30°, and the other 10°; so that in winter Quebec is usually so intensely cold as to have 22° of frost, while the south of Iceland is often without frost. In summer, on the other hand, Quebec is 15° warmer than the south of Iceland. Canada is chilled by the cold land and sea lying to the north and north-east of it. Iceland is warmed by the ocean-current (Art. 15) which, summer and winter, sweeps past its shores.

18. In order to compare the climates of two places, it is necessary to know how the temperature is distributed through the different seasons. To aid comparisons of this kind, charts are prepared like those in Plates IV. and V., showing the average distribution of temperature for each month, or for summer and winter, as well as charts constructed to show the mean temperature of each part of the globe for the whole year. From the facts embodied on such charts, gathered as they have been from all parts of the world, we conclude that (2) **temperature is regulated by the distribution of sea and land.**

19. The influence of atmospheric pressure likewise comes into play as powerfully affecting the distribution of temperature. Thus the low pressure in the region of Iceland during winter (Plate II.) causes the air to move from the south-westwards over north-western Europe, and from the north-westwards over north-eastern America. In the one case the winds come from a warm ocean and raise the temperature; in the other they travel from the cold Arctic lands and seas and therefore depress the temperature. Again, the high pressure area of Central Asia in winter is marked on the western side by a flow of warmer air from the south south-west over Russia and Western Siberia, and on the eastern side by a flow of colder air from the north-west. Thus, places on the same latitude and equally continental in position may differ as much as 40° in their mean winter temperature. We may conclude from such facts that (3) **temperature is materially affected by the prevailing winds.**

20. The form of the land likewise greatly modifies climate. Thus, a range of lofty mountains which stretches across the track of a prevailing wind, by causing a precipitation of rain (or snow), makes the air drier. Consequently, in the regions beyond the mountains, the effects of radiation are increased, the summers being hotter and the winters colder than they would otherwise be.

21. There is another way in which the form of the land influences temperature. It is a familiar fact that on low-lying land the air is warmer than on the tops of hills. Even on many of the British mountains, which are, comparatively speaking, low, the summits are so cold that in their sheltered crevices, screened from the sun and wind, snow remains unmelted through the summer. In the upper parts of the Alps, the Himalaya, the

Andes, and all the higher mountains of the globe, the cold is such that the snows of winter never wholly disappear, but remain as a perpetual covering. A gradual cooling of the air is perceptible as we rise above the sea-level in every part of the world. The rate at which this fall in temperature takes place varies much, but it is commonly taken to be on an average 1° Fahr. for every 300 feet. Within the tropics, while the low lands are lying under a burning heat, the mountains, if lofty enough to reach the upper cold air, have their summits covered with snow: hence, an elevation of a few thousand feet above the sea gives as great a change of temperature as a journey of many thousand miles away from the equator would do. From such examples we are taught that (4) **temperature is influenced by the form of the land and especially by its height above the sea.**

22. A further familiar but important fact in the distribution of temperature deserves to be noticed here. All over the globe there is a daily oscillation of temperature, the highest reading of the thermometer being reached shortly after noon, and the lowest between four and six o'clock in the morning. The time of occurrence of the maximum and minimum and the amount of difference between them vary at different places, and even at the same place, according to the season of the year. The daily range of temperature generally increases towards the equator and diminishes towards the poles; it likewise increases in proportion to distance from the sea, and is augmented by the dryness of the air.[1]

23. Since the earth is continually receiving a supply of heat from the sun, we may be tempted to ask whether it is not getting warmer. So far as observations of temperature have yet been made, they do not indicate any sensible permanent increase or diminution of heat. It would seem that the earth radiates back into space as

[1] See Buchan, Art. "Meteorology," *Encyc. Britann.*, 9th edit.

much heat as it absorbs. The quantity of heat received from the sun may be looked upon as remaining, on the whole, constant from year to year; though careful observations of the sun's surface, and especially of the recurrence of the black spots visible there, may yet show that the quantity varies from time to time, and that the variation affects the temperature and climate of our planet. It has been satisfactorily established that there is a coincidence between rainfall, including storms, and those periods when the sun's face is most marked with spots.

24. Radiation from the earth's surface is felt most at night, and especially when the sky is clear. At such times we may learn how rapidly the heat of the day is passing away from the earth into cold stellar space, and how entirely our globe depends on the sun for its present surface temperature. Objects, which in the day were warm to the touch, now grow more and more chilly. The air grows colder by contact with the cooled ground, and our own bodies, like everything else around, are at the same time radiating heat, and adding to our sensation of cold.

LESSON X.—*The Moisture of the Air.*

1. One of the constantly present ingredients in air described in Lesson VI. was water-vapour. We have found how important this component is in determining differences of pressure, and consequently in giving rise to changes of weather. It may now be considered rather more at length in reference to its sources of supply, and the different forms in which it is taken out of the air and restored again to land and sea.

2. First, then, let us ask whence does this widely-diffused and all-important vapour come? It is all **evaporated,** or given off in an invisible form, from the surface of every sea, lake, river, and spring; in short, from every water-surface on the face of the earth, and even from ice and snow. Nothing is more familiar than the rapidity with which water dries up on streets and roads after rain. Every accumulation of water, indeed, which is freely exposed to the air, and is not replenished with water, is seen to diminish, and finally to disappear. It is not that the water merely sinks into the ground. Part of it no doubt does so, but we find the water to disappear even from a saucer or other vessel where there can be no loss underneath.

3. The air is usually busy absorbing vapour. When it can hold no more it is said to be saturated, or to reach its point of saturation, and then evaporation ceases. This limit varies according to temperature, warm air, as was proved in Lesson VIII., being able to contain more vapour than cold air. Evaporation is greatly helped by wind. Damp places and pools of water, for instance, dry up in a breeze sooner than they do in still air, because the wind removes the vapour as it is formed, and brings other and drier air to drink up and carry away renewed supplies of vapour.

4. Evaporation, therefore, takes place chiefly during the day, especially the warm parts of the day, and more actively in summer than in winter. It is feeble in amount when the air is moist and still, but goes on briskly when a fresh wind blows. It takes place far more copiously in warm tropical regions than in those with a temperate or polar climate.

5. It has been calculated that the amount of water which is annually condensed out of the atmosphere upon the surface of the earth would, if collected together into one mass, cover to the depth of one mile an area of about 200,000 square miles, or a space nearly equal to the size of the whole of France. This vast liquid mass is, as it were, pumped out of the sea and out of the waters of the land by the sun's heat. But the amount of water that passes off into vapour in the atmosphere is best realised when we consider the enormous quantity

discharged into the sea by rivers. All over the world, rivers, great and small, are continually pouring their burden of water into the ocean. All the water thus supplied is obtained by the rivers from the atmosphere, either directly by rains and snows, or indirectly by springs. Vast though the amount of water is which is daily discharged into the sea by rivers, it is obviously less than the quantity actually received by the rivers themselves, for while they are coursing down from the mountains to the sea, vapour continually rises from their surfaces, and their volume consequently grows less.

6. If, then, from every ocean, lake, and river on the surface of the globe water-vapour is continually passing into the air, what becomes of all this vapour, and why do not the waters of the globe sensibly diminish? The reason is, that the conversion of water from its common visible liquid form into the invisible gaseous state is only one-half of a gigantic system of circulation. The vapour is not allowed to accumulate indefinitely in the atmosphere; it is changed back again or condensed into water, and then appears in such forms as dew, clouds, rain, hail, or snow.

7. The two processes of evaporation and condensation balance each other; that is, so far as the larger features of the earth's system are concerned, as much water is returned to the sea and land as is taken from them by the atmosphere. It is from this circulation of water that all the manifold phenomena of clouds, rain, snow, rivers, glaciers, and lakes arise. But more than this, if we reflect that now evaporation and now condensation must from time to time be predominant at any place, we perceive how greatly these processes must affect the pressure of the air; and since variations in pressure determine the various movements of the atmosphere (Lesson XI.), it is obvious how all-important this water-vapour must be in the present arrangements of our globe.

8. So far as we can tell, though the quantity of vapour may sometimes be greatly reduced, it never disappears wholly from the air at any place. On the other hand, the air round about us is comparatively seldom so saturated with moisture as not to be able to hold any more, though in damp weather we may see from the extreme slowness with which wet places dry up, how little appetite the air then has for further vapour.

9. One office of the vapour of the atmosphere is to keep the earth much warmer than it would be were the air quite dry. On the one hand, it interposes as an invisible screen between the earth and the rays of the sun, which would otherwise be intensely hot. On the other hand, the same screen, often condensing into visible form as clouds, prevents the earth at night from giving off its heat too rapidly into cold space. If it could be removed for a little from around us, we should be burnt up by day and frozen hard at night. Clouds would cease to form, rain to fall, and streams to flow, and the condition of the planet as a habitable globe would be brought to a stand.

10. When evaporation takes place, heat is abstracted by the vapour from the surface which evaporates. If you spread a drop of water over the back of the hand you feel the skin to be chilled a little, because the water in passing into vapour abstracts heat from the hand. It is a common practice to cool liquids by placing moist flannel round the vessels in which they are contained. The moisture of the flannel evaporates, and in so doing takes heat away from the vessel inside. The invisible vapour which rises into the air so abundantly carries heat along with it. This heat, not being sensible so long as the vapour remains uncondensed, has been termed *latent*.

11. Again, when condensation goes on, the heat which the vapour had held in its grasp is given out again and becomes *sensible*, as the vapour passes into water. It

has been pointed out, for example, that every pound of water which is condensed from vapour liberates heat enough to melt five pounds of cast-iron. We can well understand, therefore, that when the process of condensation takes place in nature on a large scale, the conversion of vapour back again into the state of water materially warms the air.

12. The act of condensation always occurs when air is cooled down to, or rather just below, its *dew-point* (Art. 14); but this does not always happen at the same temperature, nor does it appear in the same forms. Sometimes it issues in a light mist, or beads of dew, or drops of rain, or, if the temperature should be low enough, in hoar-frost, flakes of snow, or pellets of hail.

13. The substance which we call water is thus shown to exist in three forms, according to temperature. At ordinary temperatures, or from 32° to 212° Fahr., it is abundant as a *liquid*, and this is, of course, its most familiar condition. If at any temperature, even in the condition of ice or snow, it is allowed to stand exposed to the air, or if by having its temperature sufficiently raised, it boils, it passes off into invisible *vapour*. If again the temperature be depressed to 32°, the water begins to become *solid*. It crystallises into the brittle, colourless substance called ice. This act of crystallisation has received the name of *freezing*, and the temperature (32°) at which it takes place, is called the *freezing-point*. According therefore to the temperature at which condensation takes place, the invisible water-vapour assumes a liquid or a solid form.

14. **Dew.**—On a summer evening, if the sky is clear, condensation takes place on the leaves of plants, on stones and other objects exposed to the sky. These become covered with fine drops of water, which is known by the name of Dew. If the sky is cloudy, dew either does not form or only in small quantity, since on cloudy nights the air is not so cold as on those which are

clear. The cause of the appearance of this moisture is the same as that of the mist on a tumbler of very cold water brought into a warm room. The dew comes out of the vapour of the atmosphere, and not out of the objects on which it forms. When the sky is cloudless, radiation goes on rapidly from the earth, and the surfaces of those substances which part with their heat most readily become much colder than the air. Grass, for example, cools about twice as much as common garden-soil. As the process advances, the air next the cooling surfaces is so chilled as to be no longer able to keep all its vapour, part of which is condensed and appears as dew. Hence the grass soon becomes quite wet with the copious gathering of dew upon its blades. The temperature at which this condensation takes place is the point of saturation or **dew-point**.

15. It has been pointed out (Art. 10, 11) that the vapour of the atmosphere contains a large amount of latent heat, and that when the vapour is condensed this heat becomes sensible. The formation of a film of dew upon the surface of the earth lets heat escape back again into the air. But as radiation proceeds, the warmed air, lying on the cooling surfaces, is again chilled down to the dew-point, then more dew is abstracted, and more heat given out. In this way the nights are kept from getting colder than the temperature of the dew-point. But sometimes, when the temperature is low, as in winter, or when the dew-point is low and radiation is great, the surface of the ground is so chilled that the dew freezes in forming, and appears as the white rime, or **hoar-frost**, which we see at early morning on the grass. Clouds hinder dew from forming because they check the passage of heat from the earth into space while they themselves radiate heat towards the earth; hence it is that cloudy nights are warmer than those which are clear and starry.

16. **Mist and Fog.**—When a mass of warm and

moist air meets colder air, or comes into contact with cold ground or in any other way is cooled down below its dew-point, the excess of vapour, which it can no longer retain, condenses into minute particles, and is made visible in the form of Mist or Fog. In winter, we have a familiar illustration of this phenomenon in the condensation of our breath into mist, as it passes from the mouth into the cold air. In summer, mists frequently form in the evening over rivers and sheets of still water. As radiation goes on, the ground round the water becomes several degrees colder than the water itself, and the vapour rising from the water is consequently chilled by the cold overlying air and is condensed into banks or sheets of fog. Again, when a warm wind strikes upon a hill or mountain, and is forced to ascend the slopes, its temperature begins to fall, and if this cooling is carried below the temperature above which it can retain its vapour, that is its dew-point, the excess of vapour takes the form of a mist.

17. Clouds.—Dew and mist are formed on or near the ground, whether low plains or high mountains. Vapour carried up into the cold upper parts of the atmosphere, condenses and becomes visible there in Clouds. A cloud is only a mist hanging in the air, instead of resting on the ground. When the ground rises into the upper air, as it does in mountains, it reaches or even surmounts the layers of the atmosphere where clouds are commonly formed. We see wreaths of cloud gathering on the sides or summits of mountains, and yet if we climb up to these clouds we find them to be mere masses of mist, such as are formed down among the valleys.

18. Owing to the constant great changes in temperature and evaporation at the earth's surface, currents of air must always be somewhere ascending and carrying vapour up with them. But besides this vertical movement, the atmosphere, as far upward as can be

observed, includes many different layers or horizontal currents one above another, moving in various and even opposite directions. We may often notice the existence of these high aërial currents by watching the movements of the clouds. A lower layer of thick, woolly clouds may, for example, be rolling along, but through the rifts, a far higher layer of thin, light, white cloudlets may be drifting in a contrary direction. People who have made ascents in balloons, have found abundant evidence of the great number and various motions of these atmospheric currents.

19. When an ascending current of warm, moist air rises high above the surface of the earth, it meets with air much colder than itself, and it likewise loses sensible heat from the expansion consequent on the diminished atmospheric pressure at these elevations. Chilled down to its dew-point, it forms a cloud in the air. The cloud may be thought to remain perfectly motionless and unchanged in form, yet when watched attentively, it may be seen to alter, somewhat in the same way as a wreath or cap of mist on a mountain, and for the same reason, as explained in Art. 23.

20. It is often easy to watch how clouds grow and disappear. During summer, for instance, when the sky has been quite clear in the morning, we may observe that as the day advances white clouds make their appearance, small at first, but growing by degrees from a more or less level base into huge piles with every variety of fantastic outline. As evening comes on, these masses diminish slowly. At sunset, only a few thin flaky cloudlets are perhaps to be seen, and at last when night sets in, the sky is once more clear and cloudless. In such cases, it is to the warming of the earth by the sun during the day, and the consequent ascent of moist air-currents, that the clouds are due. Each growing mass of cloud forms the top or capital, as it were, of a column of up-streaming warm air laden with vapour. The air,

expanding and cooling as it rises, at last reaches a point where it can no longer retain its vapour, and there the cloud begins to form. After the heat of the day is over, and the warm moist air no longer streams upward, the cloud ceases to grow, and then begins to sink earthwards, as radiation goes on, until coming into warmer air, it gradually melts away and leaves the sky clear and starry.

21. As the atmosphere is traversed in all directions by currents of air of various temperatures and degrees of moisture, the meeting of these currents must often give rise to clouds, and likewise dissolve clouds already formed. A warm moist wind, for example, coming in contact with a cold wind will part with some of the vapour as cloud. On the other hand, a belt of cloud invaded by warm dry air will be evaporated and disappear. As a rule, ascending clouds increase in size, while descending clouds diminish, because in the one case they rise into colder air, and in the other, sink into warmer air. It is the continual movements of the atmosphere which produce the never-ending comings and goings of the cloud-world above us.

22. When clouds enter or are formed in one of the upper steady air-currents they are borne along, sometimes for great distances and at a great rate. On a breezy spring day, they may be seen sailing across the sky at what may seem a leisurely pace, which, however, by the rate at which their swiftly-moving shadows fly across hill and plain, is proved to be often more than 80 or even 120 miles an hour. They can be watched continually changing shape and size as they move along, rolling in huge folds over each other, sometimes lessening and sometimes increasing, and in all these movements testifying to the ceaseless turmoil of the atmosphere in which they are suspended.

23. In hilly countries, another curious feature of cloud-formation may often be seen. When a strong wind blows, sending leaves and dust high into the air, a cloud may

be observed to remain persistently on a mountain-top. Fragments of the cloud may now and then be torn off by the wind, but these, after travelling some way, gradually disappear. In such cases, the wind comes along with its warm vapour, which is quite invisible until it strikes against the colder hill-side and undergoes expansion, as it is forced up the slope into a higher level in the atmosphere. Chilled there, it passes into a fine mist, which, seen from a distance, takes the shape of a cloud

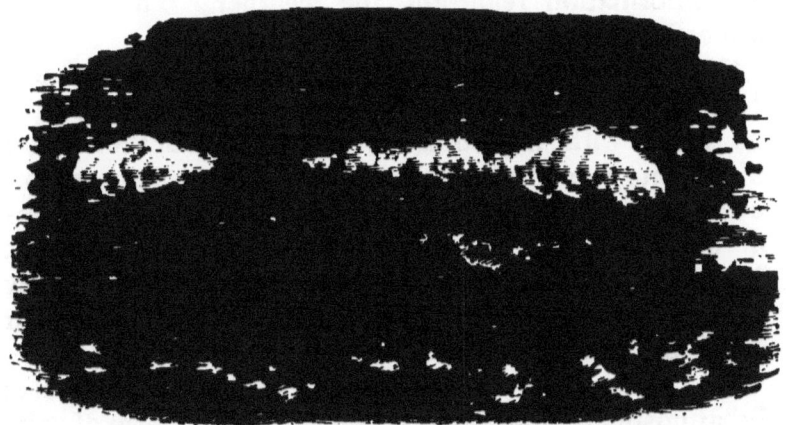

Fig. 9.—Clouds condensed by the cliff of the Noss Head, Shetland, with a clear sky and a S.E. wind. Sketched on June 14th, 1876.

capping the summit. This cloud is stationary, but the little particles of which it consists are not. The wind continues to blow up and over the mountain-top, and the vapour which it carries along is made visible as mist or cloud, while it sweeps over the chill ground. When it has passed the mountain and mixes again with the warmer air beyond, the visible cloud is redissolved and therefore melts away on the leeward side of the hill as fast, perhaps, as it formed on the windward side. The cloudlets, too, which are occasionally detached and carried away from the mountain-top by the wind, being gradually absorbed into the air again, disappear. (Fig. 9.)

24. Considerable variety exists in the forms which clouds assume, from the thin flaky cloudlets seen in the high upper air down to the huge rolling rain-clouds that descend low upon the hills, and the dull gray sheet of cloud that overspreads the whole sky. To these various forms special names have been given which, however, it is not necessary to enumerate here. Each kind of cloud is formed in particular conditions of the atmosphere, and hence a study of the clouds furnishes valuable information regarding the weather. This subject properly belongs to the study of Meteorology.

25. The chief function to be considered in the mechanism of the clouds is the way in which they supply the earth with moisture. The vast amount of water raised into the atmosphere as invisible vapour, returns to the surface of the earth, to feed the springs and rivers of the land and to replenish the sea. In this constant circulation the clouds act the part of condensers. They collect in a visible form the ever-present invisible vapour of the air, and allow it to fall back again upon the earth.

26. Rain.—By far the larger part of the vapour of the atmosphere descends to the earth in the form of rain. The little water-particles of which cloud is composed run together as the condensation proceeds. As the drops thus formed increase, they become too heavy to float in the air, and begin to fall earthwards as rain. At first they are very small, as we may feel if we happen to be on a mountain where mist is gathering into rain-cloud. But as they descend through the air they increase in size until they reach the ground in well-marked rain-drops.

27. Rain, therefore, is a further stage in the condensation of mist or cloud. If the chilling of the cloud is prolonged, rain falls from it. This may be caused in several ways. For instance, where a warm moisture-laden wind comes against a range of high mountains, and is consequently forced to ascend, it may not only be condensed into mist (Art. 19), but if cooled still more, will drop its moisture as rain. Or a cold wind which is heavy and keeps next to the ground, may wedge itself in below a warm, moist layer of air, and so chill it as to form cloud and bring down rain.

28. The fall of rain, being dependent upon the amount of evaporation, is greatest in tropical regions, where the largest supplies of vapour pass into the air, and decreases with the gradual sinking of the temperature towards the poles. But this general law is subject to some important qualifications arising from the distribution of sea and land, and the direction of the great currents of the atmosphere.

29. (1.) Although evaporation is more abundant from the surface of the sea than of the land, condensation is more active over land than over sea. Hence the rainfall is also greater on land than at sea, and over the northern hemisphere, much of which is land, than over the southern hemisphere, most of which is water.

(2.) As the ocean furnishes most of the vapour of the atmosphere, the condensation of vapour into rain upon the land is greatest near the coast-line. The sea-board of a country may be rainy while the interior is comparatively dry.

(3.) The fall of rain is greatly affected by the form of the surface of the land. Mountains act as condensers (Art. 23), and are consequently much wetter than plains. It has likewise been ascertained by rain-gauges that the amount of rain which falls at the surface of the ground is greater than at a height of even a few feet above it.

(4.) Places which lie in the path of any of the regular air-currents are wet when they cool the current, and dry when they warm it. Hence winds blowing towards the equator, since they come into warmer latitudes, are not usually wet winds; but when winds blow towards the poles, they reach colder latitudes, are chilled, and therefore become rainy.

30. Some of these laws are well illustrated in the British islands, where the rains are chiefly brought by the south-westerly winds from the Atlantic. The coast-line facing that ocean is more rainy than the east side looking to the narrow North Sea. On the western sea-board, the amount of rain which falls in a year would, if collected together, have a depth of from thirty to forty-five inches. On the east side, however, the average annual rainfall does not exceed twenty to twenty-eight inches. Where the west coast happens to be mountainous, the rainfall becomes still more copious; hence the wetness of the climate along many parts of the north-west coast of Scotland, and in the Lake district of England, where the annual rainfall ranges from eighty to one hundred and fifty, and sometimes even more than two hundred inches.

31. Great differences are observed at different points of the earth's surface in the annual amount of rainfall. Between the tropics, where rapid and copious evaporation raises a constant stream of vapour into the atmosphere, rain is heavy and frequent, so much so that this belt of the earth's surface is known as the Zone of Constant Precipitation (Lesson XI. Arts. 10, 11). Where, in this rainy region, a high mass of land lies in the path of the warm moist air-currents, the rainfall is still further increased. Thus, in India, the range of the Khasi hills stretches across the course taken by the winds called the south-west monsoons, which bring up their burden of warm vapour from the Bay of Bengal (Lesson XI. Art. 34). The result is, that the winds as they slant up the hills into the higher and cooler air have their moisture at once precipitated as rain, of which as much as 493 inches fall there in a year.

32. Again, a tract of country situated behind a high mass of land to which vapour-laden winds blow, may have little or no rain. Thus, in India, while the range of the Western Ghats, lying in the pathway of the warm

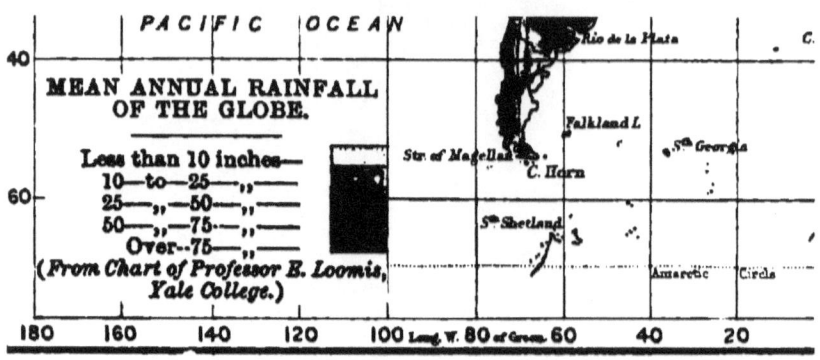

moist monsoon from the Indian Ocean, intercepts a heavy rainfall, amounting on the tops of the range to 260 inches annually, the country on the east or lee side receives comparatively little rain. Poona, for instance, lying at the foot of the hills, has a yearly rainfall of only 30 inches.

33. In South America the high chain of the Andes takes out the last of the moisture which blows from the east across the Continent, and the winds descend upon Peru so dry that rain is almost unknown there. Another and much larger rainless tract lies in the deserts that stretch from North Africa across Arabia far into the heart of Asia. In these regions, the dry, sandy soil is raised to an intense heat during the day. There is little or no water to evaporate. The hot, dry air ascends, but the winds which blow in upon the deserts cannot deposit any moisture, for instead of being cooled, they are heated and driven up in the ascending currents.

34. In some countries (Lesson XI. Art. 21), the wind blows for part of the year in one direction, and for the rest in the opposite direction. These periodical winds are generally accompanied with rain when they come from warm to cold, and with dry weather when they come from cool to warmer regions. They consequently are accompanied with rainy seasons and dry seasons. During June and July, for example, the southerly wind, referred to in Art. 31, brings the "rains" that refresh the surface of India after the scorching heat of April and May. During November, December, and January, on the other hand, a cool dry wind gently streams down from the northern mountains into the plains of Hindostan, and brings calm, dry, and settled weather. In north-western Europe, and generally in those parts of the earth which lie in temperate and arctic regions, great irregularity of rainfall occurs. In the western parts of Europe the chief rainfall is in winter, but eastwards it comes in summer.

35. Rain falls to the earth as water nearly pure. It is, indeed, natural distilled water. But nevertheless it is never quite pure, and sometimes it contains a noticeable amount of impurities. (Lesson VI. Art. 6.) Thus, it absorbs a little air, but the proportions of the atmospheric gases which it dissolves are not the same as in the atmosphere itself, there being a higher percentage of oxygen and carbonic acid than in air. Other gases and vapours are also occasionally present. All these, when dissolved in the falling rain, are brought down to the earth together with the particles of dust floating in the air. The decay of animals and plants furnishes a large supply of infinitesimal particles of organic matter, while over and above these, millions of microscopic living organisms float all through the lower parts of the atmosphere. In ordinary good air, these various impurities are, of course, in exceedingly small proportions. They are least abundant in the clear air of the mountains, and most in the bad air of towns. Now, the rain, as it were, washes these out of the air, and thus purifies it, and makes it healthier, while at the same time it carries down into the soil substances such as ammonia, which are diffused in the air, and are of service to the growth of plants. So that in addition to all the benefits which rain confers upon us by moistening and fertilising the soil, filling the rivers, replenishing the springs, and thus keeping the face of nature fresh and green, it cleanses for us the very air which we breathe.

36. **Snow.**—When from any cause and in any place, whether on the land, or in the air, water is cooled down to 32° Fahr., it usually no longer remains in the liquid form, but passes into ice (Art. 13). Ice is formed in the air by the freezing of the particles which are condensed from the water-vapour. It may descend to the ground as Snow, Sleet, or Hail, according to the circumstances in which it is formed, or the condition of the successive layers of air through which it has

to pass in its descent. Since the temperature falls in proportion as the air recedes from the surface of the earth, the freezing-point will be reached at no great height above the ground. We might, indeed, imagine a line passing through the atmosphere from pole to pole, and marking the isotherm of 32°, that is, the limit above which the water-vapour must condense into ice, and below which it will condense into water. Such a line would be liable to oscillate up and down, according to the latitude, the season of the year, and the varying currents of the air. In England, for instance, it would descend to the very ground in winter when the ponds and streams are crusted with ice during the cold period, called a **frost**, while in summer it would be a mile and a half above our heads. In India it would be three miles overhead.

37. Let us imagine, then, such a variable line or limit in the air, and consider the forms in which the frozen moisture exists there, or descends thence to the ground. It is highly probable that many of the delicate white cloudlets which we see far up in the air in summer are made of fine snow. At a lower level, flakes of snow are formed and fall earthwards, though from their feathery forms and lightness they are liable to be driven about by every feeble breath of wind.

38. A snow-flake formed in calm air will be found on examination to possess a regular structure. It is, in fact, built up of minute needles or crystals of ice symmetrically arranged into a star with six rays, each of which has a feathery surface, from the abundance of minute crystals of ice ranged along its sides. The accompanying woodcut (Fig. 10) represents different varieties of snow-flakes, each of which, however, is only a modification of the same six-rayed star. The rays strike off from each other at an angle of 60°, and no matter how complicated the form of the snow-flake may be, this angle is always maintained between all the rays.

All ice, even in the solid sheets which form on rivers and lakes, during the winter of cold climates, is built up of particles which tend to range themselves in hexagonal

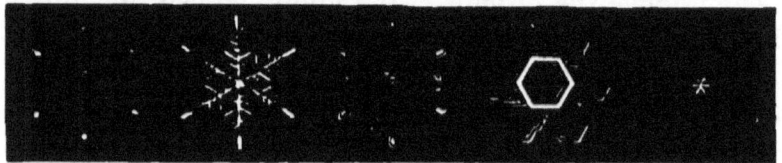

FIG. 10.—Snow-flakes.

crystals, although it is only in the snow-flake that this structure is usually seen.

39. Snow is white, but if one of the flakes be taken by itself it will be seen to be a little crystal or group of crystals of transparent ice, glittering with the prismatic colours. So that the white hue of a mass of snow arises from a combination of all these colours reflected from the immense number of minute surfaces of ice. In the same way a dish of salt appears white, and yet any one of the little crystals of which it is made up will be found to be transparent and colourless.

40. When the air is intensely cold, that is, much below the freezing-point, any snow which may then fall appears in the form of minute flakes, like a white powder. The largest flakes appear when the temperature is nearly at the freezing-point. The heaviest snow-falls do not take place during severe frosts, but before or after them. This arises from the fact that in proportion as the air grows colder it loses the capacity to retain watery vapour, so that as it becomes intensely cold it grows also comparatively dry.

41. Over by much the larger part of the globe snow never falls. It can only appear when the temperature sinks nearly to 32° at the surface of the earth. Again, it appears on those elevated parts of the land which rise into higher layers of the atmosphere, where the same low temperature prevails. Thus the Himalaya mountains,

though they stand in one of the hottest quarters of the globe, are yet so lofty that their upper portions tower far into the cold upper air, and are therefore covered with snow. On the south side of that high range the lower limit of the snow descends about 4000 feet lower than on the north side, because the cold mountain slopes condense into snow the moisture which is blown from the Indian Ocean, and allow the air to pass comparatively dry over to the north side, and because the dry air from the heated plains of Thibet evaporates the snow on the north side.

42. The **snow-line**, or limit of perpetual snow, is the

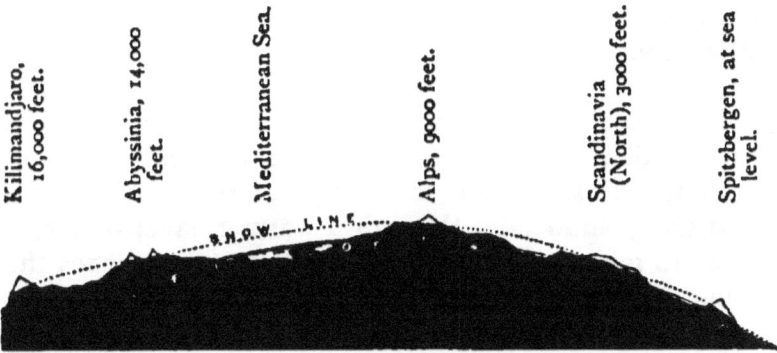

FIG. 11.—Position and height of the Snow-line between Equatorial Africa and the North Polar Seas.

name given to a line below which the summer heat suffices to melt all the snow, but above which more snow falls than the warmth of the summer months can thaw (Fig. 11). We may picture it to ourselves as a great invisible arch, of which the centre rises high over the equatorial regions while the extremities come down to the sea-level within the Arctic and Antarctic circles. Under the centre of this arch the temperature is so high that snow is only seen on the loftiest mountains, about a height of 15,000 to 20,000 feet over the sea-level, while in the far north and south where the temperature is greatly lower, the snow remains not wholly melted, even down to the margin of the sea.

43. Snow is of great use in winter, as it protects vegetation from being nipped by severe frost. Being a bad conductor of heat, it keeps the soil and plants on which it lies from parting readily with their warmth. During a frost, therefore, the soil and plants lying under a few inches of snow will be found soft and uninjured, while on those places where the snow has been blown away the soil is frozen stiff, sometimes to a depth of eighteen inches.

44. Snow that accumulates above the snow-line is pressed into ice, and creeps down into the valleys in the form of Glaciers. But this further stage in its history will be considered in a later Lesson in connection with the circulation of water over the land.

45. **Sleet.**—When snow is much driven about by the wind, its delicate fretwork of crystals is greatly broken. If this takes place when the temperature is rising, or if the driving snow falls through a warmer layer of air, it begins to melt, and in this half-melted state reaches the ground as Sleet.

46. **Hail** is the name given to pellets of snow and pieces or concretions of ice which fall from clouds. Usually **hailstones** are small, white, rounded, conical, or irregular. Sometimes, though rarely, they assume more definite crystalline shapes. They are occasionally as large as eggs, and sometimes, when several of these come together in their descent through the air, they freeze into large irregular lumps of ice. Hail is more frequent in summer than in winter, and in hot than in cool weather. Hail frequently accompanies thunderstorms, and is thus probably connected with electrical changes in the atmosphere.

47. Hailstorms are sometimes remarkably destructive. When the hailstones which fall are of large size they break branches from trees, destroy growing crops, maim

and kill cattle and human beings, and injure buildings. The course of one of these storms can thus be traced across a country by the devastation which it causes.

Lesson XI.—*The Movements of the Air.*

1. How rarely does the air seem to be perfectly motionless! Even when we say that "not a leaf is stirring," that "the chimney-smoke mounts up straight," or that "the clouds are at rest," we can find, if we look for it, proof enough that the air is not so stagnant as it seems, but is ever moving and mixing. Usually the motion is easily recognised, sometimes in a mere light air or gentle breeze, sometimes in a strong wind or boisterous gale or destructive tempest.

2. Why should the air be so restless? We have already found the answer to this question: because, owing to the unequal heating of the earth's surface by the sun, and the ever-varying amount of water-vapour poured into or withdrawn from the air, the density or pressure of the atmosphere never remains long stationary at any one place. All movements of the air arise out of differences of pressure. The law governing the direction of these movements may be stated thus:—*Air always flows in spirally from areas of high pressure to areas of low pressure.* That this must be their direction will be apparent if we reflect that low pressure indicates a deficiency, and high pressure a surplus of air. The column of air is heavier in the latter case than in the former. Consequently, obeying the universal law of gravitation, the heavier column must necessarily flow out at the base to supply the deficiency in the lighter one. The air does not flow from all sides straight into a low-pressure area. It is drawn towards and inwards upon it, and is then drawn up and passes away into higher regions of the atmosphere. This inward advance of the

air is termed a *cyclone* or cyclonic movement; while the outward flowing from a region of high pressure is called an *anticyclone* or anticyclonic movement. The rate of motion of the air is usually much higher in a cyclone than in an anticyclone. Storms of wind and rain are associated with cyclones; while calm air or light breezes accompany anticyclones.

3. We are accustomed to think only of the horizontal movement of the wind along the surface of the earth. But to form a proper idea of the circulation of the air, we must look upon that horizontal movement as only one part of a much wider system. Each cyclone or area of low pressure has at its centre an ascending current of air; each anticyclone or area of high pressure has a descending current, and it is the position and extent of these upward and downward currents which govern the direction of the winds at the surface.

4. The rate of motion of the surface-currents or winds must evidently be determined in each case mainly by the amount of difference in the pressure, and by the distance between the centres of the two areas of higher and lower pressure. The greater the difference of pressure and the shorter the distance within which it occurs, the more rapid will be the flow of air. This is referred to again in connection with storms in Arts. 24-29.

5. As there is so constant a relation between atmospheric pressure and atmospheric movement, a knowledge of the distribution of pressure over the globe will supply the clue to the disposition of all the great movements of the atmosphere. When the average readings of the barometer over a continent or over the whole surface of the globe for any given period are known, it is easy to tell what must be the direction of the chief aërial movements at any particular place. Plates II. and III. illustrate this relation between the distribution of atmospheric pressure over the globe and the general circulation of the atmosphere.

6. The principal causes of differences in atmospheric pressure were stated, in Lesson VIII. Art. 18, to be temperature and water-vapour. There seems to be at present no method of ascertaining how far any particular movement of the air is due to the one or the other of these causes, or to both combined. Directly or indirectly, indeed, everything might be referred to temperature, for evidently the process of evaporation, on which the addition of water-vapour to the air depends, is mainly regulated by temperature, being greatest when the temperature is high, and least when it is low. But it is useful to distinguish between the effect produced by the mere heating and cooling of the air and that due to the changes in the amount and distribution of its vapour.

7. Let us take some simple illustrations of the influence of temperature in producing movements in the air. A fire, whether in a room or out of doors, furnishes an excellent example on a small scale of what takes place in nature. When a fire is kindled in a grate, the air overlying it is heated, and ascends; but at the same time there is an in-draught of air at the bottom of the grate. A circulation is established. The air from all parts of the room and from outside, through the crevices of the doors and windows, is drawn towards the fire, warmed, and driven up the chimney. If in any way this free circulation is interrupted, the fire does not burn well, while if it is stopped, the fire goes out. Again, when a house or any large building, or a prairie or forest, takes fire, the heat may be so great as to cause the rapid ascent of a considerable mass of air above the burning materials. This is necessarily accompanied by a flow of air inwards from all sides along the ground towards the fire as a centre, and this in-draught may amount even to the rapidity and force of a violent wind. Rushing round towards the blazing centre, this wind feeds the flames, and, as it gets heated, ascends with great vigour, carrying clouds of smoke, sparks, and even, it may be, large fragments of burning wood along with it.

8. Though the sun's heat is not concentrated upon a limited spot like a burning building, or a forest on fire, yet the heating of a mass of land during the day produces similar effects, where the heated land lies near some other area (of water, for example) which has not been heated so much.

Nowhere can this be more instructively seen than by the seaside, when the days are warm and the nights cool. During the day, the land surface becomes much warmer under the sun's rays than the sea does, and the air resting on it becomes warmer than that resting upon the sea. The result of this difference of temperature (and therefore of pressure) is to set in motion a light breeze, which moves from the sea to the land, in order to take the place of the hot air that is always streaming up from the heated surface of the land. This is a **sea-breeze**. It dies away towards evening. When the nightly radiation begins, however, the land parts with its heat more readily than the sea does, and consequently cools the air more. The cooler and denser air above the land moves seaward to take the place of the warmer and lighter air that rises from the sea. Hence arises a **land-breeze**, which, increasing in force, as the difference of temperature between sea and land is augmented by continued radiation, may rise to a good stiff breeze.

In mountainous countries, where the higher ground rises far up into the colder layers of the atmosphere, another beautiful illustration of these changes of movement in the air may be watched. During the day the air, warmed on the bare mountain-sides, ascends, and a breeze blows up the valleys towards the heights. At evening, when the mountain-slopes begin to cool rapidly, the cold heavy air lying on them flows down as a cool breeze into the valleys.

9. As regards the influence of the water-vapour on movements of the atmosphere, it may be expected to be most marked where the temperature is greatest. The broad belt of low atmospheric pressure between the tropics (Lesson VIII. Art. 19), should be the part of the earth's surface where this influence may be best studied. There the sun's heat is greatest and the evaporation most abundant. The tropical belt, indeed, with its wide ocean surfaces, may be looked upon as the great evaporating cauldron of the globe, whence comes most of the moisture that is distributed by the winds as rain and snow. Air, from north to south, must be constantly pouring into it, because it is a vast area of low pressure. Were there no interrupting masses of land the regularity of this inflow of air would be continuous all round the globe. A constant wind would blow towards the equator from either side, and between the two opposite currents there would be a zone, in which the currents would meet and ascend as an upstreaming mass of air, along the centre of the tract of low pressure. The position of this zone would vary according to the position of the sun. In July it would be about the position of the Tropic of Cancer. Thence, following the sun in his southward passage, it would travel to the line of the Equator in October, and to the Tropic of Capricorn in January.

By the present arrangement of land and water, however, this regularity is greatly interfered with. It is best shown in the wide Pacific Ocean, where there is least land. Also, it is more distinct in January than in July, because the belt of low pressure then lies nearly along the line of the equator with more sea in its track than in July, when it goes north of the equator, spreading over Asia and the northern part of Africa.

10. Putting aside for the present these irregularities, let us think of the air round the tropical belt of the globe as constantly receiving an enormous volume of water-vapour from the oceans. From this great influx of vapour, as well as from the high temperature along that belt, there is a continued ascent of heated and humid

air into the higher regions of the atmosphere, and a consequent in-draught from north and south. This tract of the earth's surface, extending in the Atlantic region from lat. 11° N. in August to lat. 1° N. in February, and having a breadth of from three to eight degrees of latitude, is called the **Belt of Equatorial Calms**—a somewhat misleading name, because, although there are no constant winds, the air is in ceaseless disturbance, as it is heated and streams upward into the cloudy sky.

11. Constant Winds and Aërial Currents.—The hot moist air that mounts up from the equatorial belt expands and cools as it ascends, and in so doing parts with much of its moisture, which condenses and falls as rain. Hence the atmospheric pressure is much reduced. Moreover, the belt is distinguished not only by its great heat, but, in many parts, by its almost continual rains and thunderstorms, so as to be spoken of as the Zone of Constant Precipitation (Lesson X. Art. 31). Reaching higher parts of the atmosphere the equatorial hot air divides, one part flowing northward, the other southward, and each as an upper current taking a direction opposite to that of the in-draught at the surface. These upper currents must be many thousand feet above the sea at the equatorial belt. They travel pole-wards until, entering the belt of high pressure they descend, and at last reach the surface in temperate latitudes. Thence the air streams again outward, part of it back along the surface to the equator and another portion onwards into low pressure areas in higher latitudes.

12. Here, then, we have the fundamental aërial circulation of the globe—a great system of (1) surface currents, continually streaming out of the areas of high atmospheric pressure towards those of low pressure, and of (2) upper currents, constantly flowing away from above the low pressure areas. Considerable difference is observable in the distribution of the winds according to the

season of the year (Plates II., III.), and this difference is manifestly dependant on the shifting of the areas of high and low pressure. In winter, for example, the winds of the northern hemisphere flow out of the great centres of high pressure that lie over Asia, the western territories of America, and a tract of the Atlantic westwards from Northern Africa, and stream into the low pressure equatorial belt on the one side and into the low pressure tracts of the North Atlantic and North Pacific on the other. In summer these directions are on the whole reversed, because the high pressure areas have in great part become areas of low pressure.

13. It might be supposed that currents flowing to and from the equator should have a direct north and south course. And so they would, were it not that the earth, instead of being at rest, is always rotating on its axis. Owing to the motion of rotation, an object at the equator is carried along with a higher speed than an object nearer the pole, just as the rim of a wheel moves through more space in the same interval of time than the parts near the axle. Hence, when a current of air travels away from the equator northward or southward, it moves from a region where the velocity of the surface is greater to one where it is less. At the equator the air, having the same rate of rotation as the rest of the surface, is borne along with the rest of that surface, and if it blows eastward or westward, rotation does not affect its direction. But when it blows northward or southward from the equatorial tracts, it carries with it part of its equatorial rapidity of rotation, and moves a little faster than the surfaces over which it successively moves. Now, as the earth rotates from west to east, all the air that flows from the equator towards the pole is bent towards the east. On the other hand, as the air which travels from higher latitudes towards the equator is always getting into regions where the speed of rotation increases, it seems to lag behind,

and instead of flowing straight north and south, is deflected towards the west.

14. The lower or surface currents, turned more and more westwards as they advance, become on the north side of the equatorial belt the **North-east Trade-winds**, and on the south side the **South-east Trade-winds**. These are constant winds. They always stream towards the equator because of the lower atmospheric pressure there. They received the name of "Trade-winds" in the early days of navigation, when it was found that they blew so steadily in one direction that they could always be reckoned on for the purposes of commerce.

The Trade-winds are chiefly felt between each tropic and the equator, but they begin in some places considerably beyond the tropics (Art. 10), their constancy being better exhibited over the oceans than over the land, seeing that the more marked heating and cooling of the land by day and night (Art. 8) tend to disturb their regularity. Beyond the limits of the Trade-wind region, along the belt of high atmospheric pressure, there is in each hemisphere a belt of calms and variable winds, that on the north side being called the Calms of Cancer, that on the south side the Calms of Capricorn. These Calms, unlike those of the equatorial belt of low atmospheric pressure, are characterised by generally bright weather.

15. Out of each high pressure belt the winds blow towards the equator on the one side and towards the pole on the other. As the direction of the Trade-winds is deflected to the west by the influence of the earth's rotation, so the winds which emerge from the polar side of the tropical calms acquire an easterly direction. In the northern hemisphere therefore they come from the south-west. In Britain, and the west of Europe, which lie considerably to the north of the Calms of Cancer, the prevalent winds are the familiar south-west winds. In these countries the large towns commonly grow towards

the west, and the term "west-end" usually means the quarter where the newest and best streets or dwelling-houses are situated. The reason is evidently to be sought in the direction of the prevalent winds, for the west and south-west are the quarters that escape most of the town-smoke, which is blown eastward.

16. Much interesting information may often be gathered respecting the movements of the high upper currents of the atmosphere, as well as of lower aërial currents, by watching the clouds. Fine light fleecy clouds may be noticed at a great height, sometimes with other heavier clouds lying below them. They may be seen to move and change their shape, affording by these movements valuable prognostications of coming weather.

17. But the fact that there are upper currents in the atmosphere is now and then brought home to us still more strikingly. Volcanoes have been already referred to as mountains from which melted rock, steam, and gas are given out, and from which fine dust is sometimes thrown for an enormous height up into the air. With such force is this dust or ash ejected, that it sometimes enters a high aërial current blowing steadily and strongly in one direction. It is then borne along to vast distances, and descends to the earth, perhaps in regions where nobody ever saw such a thing as a volcano. In this way, the inhabitants of the northern parts of the British Islands have more than once been visited with showers of dust that came from Iceland. In the year 1783, during a great eruption of Skaptâr Jökul, one of the volcanoes of that island, the fine impalpable dust fell in such quantities between the Orkney and Shetland Islands during north-west winds, that vessels sailing there required to have the deposit shovelled off their decks every morning. In that season, too, the crops along the north coast of Caithness proved a failure, from the quantity of volcanic material which fell on the ground. The inhabitants still speak of the year of "the ashie."

The distance from Iceland to the coast of Caithness is upwards of 600 miles. Similar volcanic dust has been frequently carried from the Icelandic volcanoes to Scandinavia.

18. Instances are on record where the erupted ashes have been borne over a much greater space. In the year 1835, Coseguina, a volcano in Guatemala, broke out in eruption, and though the regular east wind was then blowing, the fine ashes were carried eastward by an upper current, and fell on the island of Jamaica, 800 miles from the point of emission. They had been four days on the journey, and their rate of progress must have been 200 miles a day. Again, in 1815, during a calamitous volcanic outbreak in the island of Sumbawa, lying to the east of Java, the ash was carried for 800 miles to the east, as far as the islands of Amboyna and Banda, although the south-east wind was then at its height.

19. In countries bordering the Mediterranean, and even as far west as the Cape de Verde and Canary Islands, the air is occasionally filled with a peculiar reddish, or brown dust, which sometimes falls in such quantities as to cover the decks of ships at sea, though far out of sight of land. When rain falls at the same time it brings down this dust with it, and having then a peculiar colour, is known as "blood-rain." It seems that in most cases this "red-fog," "sea-dust," or "sirocco-dust," as it is called, has come from the hot deserts of Africa, whence, swept up by whirlwinds from the burning soil in great clouds, it has been borne away by the overflow of air at the top of the hot ascending column, and after being carried for hundreds of miles by the upper currents, has at last descended to the surface again.

20. Clouds, ashes, and dust, therefore, sometimes give important evidence as to the direction in which the upper aërial currents are moving, and help to prove how regular and constant is the circulation of the atmosphere.

21. Seasonal or Periodic Winds.—When the two charts of the world in Plates II. and III. are compared, it may be seen that the larger masses of land in the northern hemisphere interfere a good deal with that regular distribution which, as shown by the southern hemisphere, is promoted by a broad unbroken expanse of ocean. In January, for instance, the high and cold table-lands of Central Asia become the centre of a vast area over which the pressure of the air is high. Consequently from that elevated region the wind issues on all sides. In China and Japan it appears as a north-west wind. In Hindostan it comes from the north-east. In the Mediterranean it blows from the east and south-east. But in July matters are reversed, for then the centre of Asia, heated by the hot summer sun, becomes part of a vast region of low-pressure, which includes the north-eastern half of Africa and the east of Europe. Into that enormous basin the air pours from every side. Along the coasts of Siberia and Scandinavia it comes from the north. From China, round the south of the continent to the Red Sea, it comes from the Indian Ocean, that is, from south-east, south, or south-west. Across Europe it flows from the westward. Hence, according to the position of any place with reference to the larger masses of sea and land, the prevalent direction of its winds may be estimated.

22. On the shores of the Indian Ocean the summer and winter winds are known as **Monsoons**—an Arabic word signifying any part or season of the year—but now generally applied to all winds which have a markedly seasonal character. Since the air is drawn in towards the heart of Asia in summer, and comes out from that centre in winter, the direction of the monsoon at any place depends upon geographical position. In India the winter wind is the N.E. Monsoon, which corresponds to the N.E. Trades of the North Atlantic and North Pacific Oceans; the summer wind is the S.W. Monsoon, which is a complete reversal of the natural course of the Trade-wind owing to the enormous in-draught caused by the low summer pressure over Asia. On the Chinese coast the winter wind is a N.W. monsoon, and the summer wind a S.E. monsoon. Similar but not quite so strongly contrasted monsoons occur in North America. In the Southern States, for instance, the winter wind comes from the north-east, the summer wind from the south-west. Even in Euorpe the prevalent winds partake of the character of Monsoons. During winter, when the atmospheric pressure is low in the north and north-west, they blow from the land into the North Atlantic and Arctic Oceans. In summer, on the other hand, when the low pressure is shifted to the great Asiatic land, the winds blow from the sea, as the prevalent west and south-west winds, bringing the moisture of the Atlantic to condense it in rain-showers as they pass eastwards.

23. **Local Winds.**—Many winds, often of a destructive character, occur in different countries or in different districts of the same country, to which local names are given. When they come from tracts where the pressure is high and the temperature low to where the pressure is lower and the temperature higher, they are felt as cold blasts, whereby the humidity of the air in the low pressure area is condensed into torrents of rain. One of the best known examples of such a wind is that known as the *Mistral*, which descends from the high plateaux and plains of Central and Eastern France, and is felt as a cold and sometimes tempestuous wind along the shores of the Mediterranean. When a low atmospheric pressure happens to arise on the borders of a hot desert region like those of Africa, Arabia, or the interior of Australia, it draws in towards it the hot air lying over the burning sands, which in the countries where it blows is extremely unhealthy. In Italy such a wind is known as the *Sirocco*—a hot moist wind which raises a haze in the air, and produces a sensation of

extreme languor both in man and beast. In Spain, where it receives the name of the *Solano*, it sometimes comes across the narrow part of the Mediterranean laden with fine hot dust from the vast African deserts. In Africa and Arabia it appears as the dreaded *Simoom*—a hot suffocating wind, which sometimes rushes across the desert with such violence as to raise clouds of sand, and sweep them in whirling masses for many miles. It thus heaps up vast mounds of sand, under which caravans of travellers may be completely buried. One of the armies of Cambyses, 50,000 in number, is said to have been engulphed in the sand when on its way to attack the oasis and temple of Jupiter Ammon. Again on the coast of Guinea, during December, January, and February, a hot wind, called the *Harmattan*, blows from the interior out to sea. The north-west provinces of India have likewise their hot-winds, which sometimes produce violent whirlwinds sweeping up the dust and carrying it in tall whirling columns into the upper air, whence it gradually finds its way to the earth again.

24. Storms.—But besides the winds and currents already noticed, most of which blow continuously for weeks or months, sudden and violent commotions arise in the atmosphere, and are often very disastrous in their effects. These may be classed under the general name of Storms. Let us see whether they can be accounted for by the general principle of aërial circulation, which holds good for the constant and periodic winds.

25. By careful observation of the barometer, and more particularly by having observing stations at many different places from which the readings of the barometer can be telegraphed at frequent intervals to some central office, it is possible to forecast the weather and to detect the approach of storms before they actually arrive. A system of weather-forecasts has now been in operation for some years in Western Europe and in the United States. When a violent storm bursts upon a country the wind rises rapidly into a gale and continues to rush along at a rate of often 70 or 80, sometimes in gusts, even of 120 and 150 miles an hour. After some hours it slackens its speed, and may even die away as rapidly as it rose. But again it may spring up from a different or even from the opposite quarter and possibly equal, if it does not exceed, its first fury and destructiveness. All parts of a country have not the storm simultaneously. In Europe, for instance, storms usually come from the westward. Britain and the western shores of France and Portugal catch the first fury of the tempest, which then travels eastward or north-eastward, and not unfrequently dies away in Russia. The rate at which the storm moves from place to place does not exceed on an average from fifteen to thirty miles an hour—a much more leisurely rate than the rapid and destructive rush of the wind would have led any one to expect.

26. Suppose that such a storm has passed over Europe, and that after it is gone, we obtain information as to the readings of the barometer over all the countries traversed by the storm as well as those around its track. We find that the centre of the storm has been an area of very low atmospheric pressure, 600 miles or more in breadth, and that the pressure has been much higher immediately outside of that area. The great law of atmospheric movement, therefore, holds good here also. The air has moved as a cyclone inward from surrounding regions of high pressure upon a central area of low pressure.

27. Observations of this kind have proved, that the greater the amount of difference between the low pressure in the centre of the storm and that of the high pressure outside, and the nearer the extremes of pressure lie to each other, the more violent is the wind. Suppose, for example, that while the barometer stood at 30 inches at one side of a country, it should rapidly fall to 28 inches at some district on the other side, 300 miles away: so great a difference in so short a space would certainly be accompanied by a violent gale. The air in such circumstances rushes in with an in-moving spiral and ascending motion. It turns inwards upon the storm-ring, and is carried upward in the central vortex as a vast up-streaming current, which overflows at the top and passes off into other regions.

28. Hence it is evident why, though the wind rushes so furiously, the whole body of the storm does not travel so fast in many cases as an ordinary passenger train. There are two motions in the storm, that of the whirling air as it is borne inwards, and that of the whole rotating mass or storm. When the storm-ring bursts upon a place the wind may be blowing from the south-east. It dies down when the centre of the storm reaches the place, but rises again from a different quarter when the remaining side of the storm-ring advances, and continues to blow until the whole whirling mass of air has gone by.

29. A good illustration of this vorticose movement of a storm is afforded by the little dust-whirlwinds often seen upon a dusty road in dry weather. The air in each of these, as shown by the movements of the dust, takes a rapid inflowing spiral course, drawing in air on all sides below and whirling it rapidly round and upward, till it expands at the top and flows over into the surrounding air.

30. **Office of the Winds.**—There are two kinds of work done by the winds and currents of the atmosphere (1) the distribution of temperature; and (2) the distribution of moisture. This work has been referred to already, but may again be stated in a summary way here.

31. (1.) When a wind blows from a warm or mild quarter, it raises the temperature of the places to which it comes. The south-westerly winds of Britain, for instance, warmed by the heated waters of the North Atlantic, keep the air of that part of Europe much milder than

from its latitude it ought to be. When, on the other hand, a wind blows from a cold to a warm region it lowers the temperature as it moves along. Thus, owing to the vast expanse of cold land in Asia during winter and the high pressure there, the outflowing winds are cold. A careful study of the charts in Plates II. and III. will show this more clearly than a description in words can do. By noting the areas of low and high pressure, we learn in which main direction the wind must blow in any part of the globe; then a reference to the isothermal charts will indicate to us the temperature of the high pressure tracts from which the wind comes, and therefore whether the wind in any particular case must be cold or warm.

32. (2.) The winds are the great agents by which the moisture of the atmosphere is distributed over the globe. Were it not for their help, the condensed vapour would be discharged over the same tracts from which it had been evaporated. And therefore, as by far the most of the vapour rises from the ocean, it would fall back again into the ocean, and the land would only get back the comparatively small quantity evaporated from rivers and lakes. But since the winds bear the vapour along with them, and since areas of low atmospheric pressure often pass over continents, the vapour is carried away from the sea, and condensed into rain over the land. On the whole, therefore, the wetness or dryness of any place will depend on the direction from which its prevalent winds come (Lesson X. Art. 29-34). If they blow from a wide and warm expanse of sea they will be laden with moisture, and be ready to drop it when chilled by passing over the land. If, on the other hand, they blow from the interior of a continent, they will be dry—their heat or cold depending upon the temperature of the region whence they come.

33. In short, winds are wet when they travel from a warm vapour-laden tract to a colder one, because they are cooled below their dew-point and must let fall the surplus moisture. Winds are dry when they travel from a cold to a warm tract, because, instead of parting with their vapour, they are ready to take up more; they are dry, too, when they issue from a hot and dry region, and before they have an opportunity of crossing any sea and greedily drinking up vapour from its surface.

34. When a mass of high land stretches across the track of a warm moist wind, it forces the wind to flow up and over its ridges. The air is thereby expanded and cooled, and as a consequence deluges of rain descend. The most notable instance of this kind is that of the Khasi Hills already mentioned (Lesson X. Art. 31), which front the Bay of Bengal, and interpose as a great steep buttress against the inland advance of the warm vapour-laden south-west monsoon. Driven up the slopes, the wind has to discharge its moisture in rain, of which, during the seven months, when that monsoon blows, a depth of as much as 493 inches, or about 41 feet, falls to the ground. But, after passing over the hills, the wind, having lost so much of its moisture, is comparatively dry. Hence the rainfall rapidly diminishes northward. As the wind journeys in that direction, however, it is once more forced to ascend into higher regions of the atmosphere by the giant chain of the Himalaya. Its moisture is further wrung out of it, as rain on the lower slopes, and as snow among the higher ridges. And thus it descends on the north side as a cool dry wind into the wide plains of Thibet.

35. In countries like India, therefore, the dry season and rainy season depend upon the changes of the monsoon and the quarter whence the monsoon blows, whether dry land or vapour-giving sea.

36. In regions where the winds are irregular the rains are found to be the same. In the west of Europe, for example, the wind shifts continually, but the prevalent winds are from the west and south-west, though during part of the year east winds become frequent. The westerly winds come from the warm Atlantic, laden with vapour, which they discharge plentifully on the western parts of Britain and the Continent, leaving the eastern districts comparatively dry. The easterly winds on the other hand come from the heart of the Old World, and are so dry as often to wither up vegetation as much as a hard frost or even a scorching fire would do.

37. There is yet another kind of work performed by the winds, which, though of very limited and local character

FIG. 12.—Sand-dunes—ridges of dry sand blown inland off the shore by the wind.

compared with their other services, deserves notice here. Storms by sweeping over the surface of the land sometimes produce great changes there, uprooting trees and even prostrating large woods, destroying fields and houses, and generally carrying ruin with them in their track. But even where the wind does not rush with such fury over the land it sometimes effects remarkable changes there when its prevalent direction is from the sea to the shore. On sandy beaches exposed to winds which usually blow from the sea, the dried sand is often driven inland, where it forms mounds and ridges called sand-dunes, sometimes 60 or 100 feet high, fronting the beach for, it may be, many miles. A small streamlet, flowing behind these ridges may arrest their advance by carrying away the loose sand as fast as it creeps on. But where no such check exists, the sand may gradually extend inland, and completely bury the cultivated soil over whole farms or parishes.

38. Many examples of these effects are to be found along the coasts of the British Islands. Within the last few hundred years, thousands of acres of valuable land have been overwhelmed by the drifting of sand from the sea. Along the shores of the Moray Firth and of Aberdeenshire, for example, several parishes have been wholly or in great part buried. On the coasts of Norfolk and Suffolk, villages and thousands of acres of land have been covered with blown sand during the last two centuries, and on the Cornish coast similar inroads have taken place. The shores of the Bay of Biscay furnish still more striking instances of the power of the winds to alter the surface of a district. The dunes which are there heaped up by the westerly gales, march inland at a rate of 60 or 70 feet in a year. No barrier, natural or artificial, is able to withstand their progress. Fields, woods, and villages are buried in succession. Nor is this all. The sand ridges interrupt the drainage from the interior, and the water collects among and in front of the ridges. Ponds and lakes are formed, which, unable to find an exit, are driven inland along with the sand barriers which dam them up. Large tracts are thus first inundated with water, and then finally overwhelmed under the advancing sand. Roman Roads, and many villages which existed in the Middle Ages, have disappeared. And the same destruction is going on still.

39. It is not always along the sea-margin that these effects of the action of wind are to be traced. In the interior of many countries there are wide tracts of barren sand or Deserts (Art. 23), over which the wind raises the sand into ridges, as it does along the shore. Such, for instance, are the great desert of Sahara and much of the interior of Arabia.

40. In dry climates vast quantities of dust are transported by wind. Thus in Central Asia, the air is often thick with a fine yellow dust, the sun being sometimes so obscure at mid-day that lamplight is required to read. The dust settles down as a fine sediment, covering everything and increasing the soil. In some countries, ancient cities and ruins have been gradually more or less buried under the gradual accumulation of wind-borne dust, aided by vegetation which intercepts the sediment and grows over it. Nineveh, Babylon, and other historic sites appear to have been covered up in this manner.

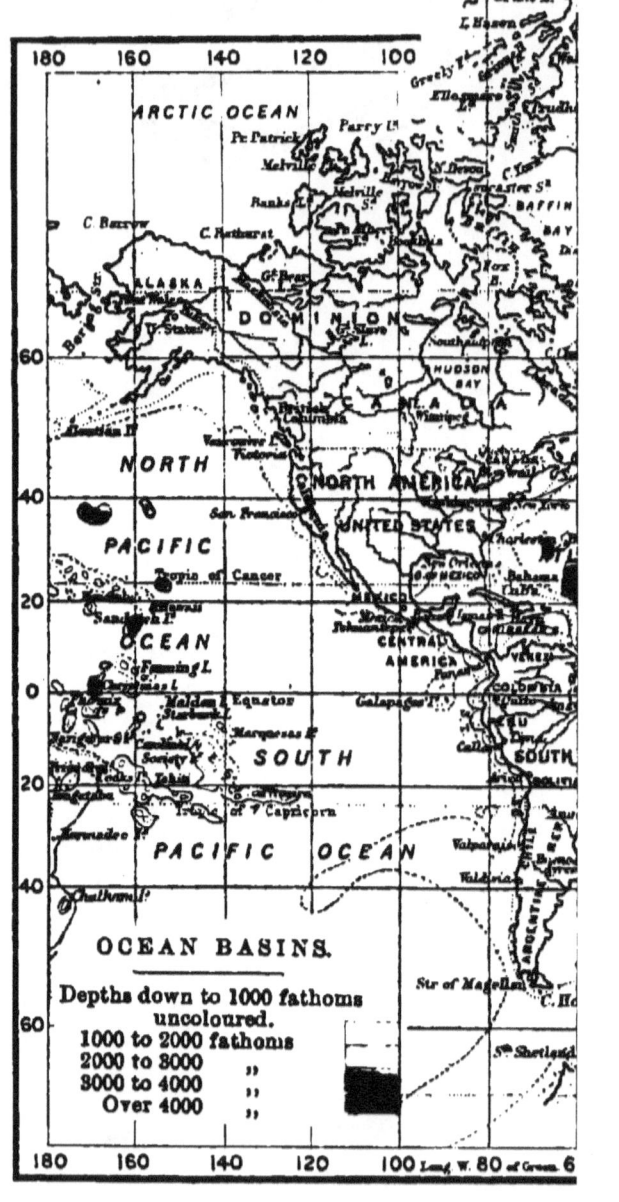

CHAPTER III.

THE SEA.

Lesson XII.—*The Great Sea-basins.*

1. From the outer envelope of air which encloses the earth we now pass to the underlying envelope of water called the Sea. At the outset some obvious differences between these two coverings may be noticed. For example, while the atmosphere completely wraps round the whole planet, rising to a height of many miles above its general surface, the water envelope is pierced in many places by masses of the underlying solid part of the earth, which rise above it to form land. As already mentioned (Lesson V. Art. 2), the sea covers not quite three quarters, and the land a little more than one quarter, of the entire surface of the earth.

2. Again, we know nothing about the upper surface of the atmosphere, and cannot say precisely how far it lies above us, but the surface of the sea forms a great plain, and the line between it and the air is sharply defined. Though we speak of the sea as forming a plain, we know this apparent plain to be really curved, and that from its wide extent and its freedom from inequalities, it shows the curvature of the earth's surface better than can be seen on land. (Lesson I. 2.) The amount of the curvature is about eight inches in a statute mile; that is, an object eight inches high above the sea-level sinks out of sight when looked at from the same level at a distance of more than a mile. The line of meeting between the

sky and the surface of the earth is termed the **horizon**. Its distance from us evidently depends upon the elevation at which we may happen to stand. Thus, at the sea-shore with our eyes exactly six feet above the sea-level, our horizon out to sea is three miles off. If we ascend so that our eyes are about ten feet and a half above the sea-level, our horizon is extended to four miles. If we climb to some adjoining height, say to a lighthouse top, about ninety-six feet above the sea, our horizon is increased to a distance of twelve miles.[1]

3. Another evident contrast between the air and the sea lies in the fact that while every inhabitant of the earth is familiar with the one, only a comparatively small part of mankind has ever seen the other. Even in an island like Britain, a large proportion of the inland population has never been within sight of the sea. On the continents the proportion is necessarily much greater, for, except the people dwelling along the sea-margin, the great mass of the inhabitants, having little or no communication with the coasts, have no acquaintance with any larger sheet of water than their own native river or lake.

4. One who has not seen it can hardly realise what the sea is from descriptions in books. Let us suppose, however, that some intelligent dweller in the inland regions were taken for the first time to the sea-coast, and, after recovering from his first impression of wonder and admiration, were to begin to look attentively at those features which would be most likely to attract his notice. He would observe that the solid ground, with which he had been familiar all his life, gives place to a seemingly boundless plain of water, at first sight level and motion-

---

[1] In estimating the extent of the visible horizon, however, when the distance exceeds half a mile, we need to take into account the effect of atmospheric refraction which tends to make distant objects seem higher than they are. The allowance to be made for this effect varies from day to day; it commonly requires a deduction of about one-seventh from the apparent height of an object.

less, but soon found to be in a state of perpetual unrest, and answering to all the movements of the air above, heaving or rippling when the air is gently stirred, but rising into waves and foam-crested breakers along the shore when the wind blows strongly. Should he taste some of this blue sparkling water he would find it salt and undrinkable, even though clearer perhaps than his own river at home. Gathering the shells and other remains of the living things of the sea from the sands of the shore, he would find every one of them different from anything he had ever found on the land or in fresh water. Were he to watch by the shore from day to day, he would notice that twice every day the water advances slowly and as slowly retreats, and that this regular movement takes place whether the water be smooth or rough. Were he to set sail upon the seemingly boundless plain of water, he would watch the land behind him gradually sinking, as it were, into the water, till at last its highest point had disappeared, and the same long level line of meeting between sea and sky would then sweep around him on every side. With no land in sight and no other vessel perhaps to be descried; with only the sky overhead and the heaving water beneath and around, he would learn better than from any map or description the vastness and solitude of the great deep. And yet from the deck, or even from the mast-head of his ship, only a comparatively small patch of the sea could be seen at once. The horizon or sky-line, which he thinks so immeasurably distant, is only a few miles off (Art. 2). And he might sail for weeks together, passing over thousands of miles, with all the time the same limited horizon and the same monotony of sea and sky.

5. The traveller who finds himself for the first time on the open sea is not only impressed by the vastness of its surface, but by a vague notion of its profound depth. He has read of the "unfathomable abysses of ocean," and naturally feels some awe at the thought that they

are actually lying beneath him. In recent years, however, Expeditions have been fitted out to measure the depth of the sea in all parts of the globe, and the result has been to show that even the deepest of the so-called abysses probably does not much exceed five miles, while the average depth of the sea may be taken at about half that amount.

6. Had the earth been a perfectly smooth homogeneous globe, like a ball of polished steel, we may suppose that the sea would have completely covered its surface. There would then have been an outer shell of air and an inner shell of water, within which the solid planet itself would have lain. The average depth of such an universal sheet of water, as may be inferred from what has been made known by numerous soundings in all parts of the present sea, would have been about a mile and a half. It is almost certain, however, that the water envelope never completely wrapped the globe round in this way.

7. Looking at a map of the world, we may be disposed to ask whether any reason can be given for the shape which the great ocean-basins have taken; whether, for example, the sea, by the force of its own waves and currents, could have hollowed out the great depressions in which it now lies. That the sea could have had little to do with the formation of its basins will be evident if we reflect that it is only the upper parts of the sea which eat away the solid part of the earth (see Lesson XVIII.); the water in the deep abysses does not even stir the fine mud, which, like dust in a deserted room, slowly settles down upon the bottom. Where the waves wear down the land, the materials which they remove are not destroyed, but are carried away into some still part of the sea, where they accumulate on the sea-floor. So that even if we could suppose it possible that the sea might have scooped its basins out of the solid surface of the globe, it could only remove the debris from

one part of its bed to another, and, in short, fill up its hollows after it had formed them.

8. Two causes may be assigned for the present distribution of the oceans. In the first place there must be an excess of density in the southern hemisphere, so that the larger mass of water is attracted to that half of the globe. In the second place, the surface of the solid globe has been irregularly depressed and ridged up. Consider what was said in Lesson III. about the probable history of the earth. If our planet has slowly cooled from a condition of intense heat, it must have gradually contracted in bulk. We may compare it to an apple which is kept a long while, until it begins to dry up, and consequently to shrink. The apple does not diminish equally all round. Some parts of the skin sink down, others are puckered into wrinkles and foldings, so that in the end the once smooth surface becomes shrivelled and uneven. In the case of the earth, the contraction has likewise been irregular. Some portions have sunk down, others which shrank up soonest have become more and more wrinkled, and rise far above the rest. The depressed parts have formed basins for the oceanic waters. Those portions of the larger wrinkles which rise above the sea-level form continents and islands.

9. Until a comparatively few years ago, the abysses of the sea were as unknown as the interior of the earth. Their depth was only guessed at, for the few attempts to measure it had led to great exaggeration. The nature of their floor was equally a matter of conjecture. It was supposed, indeed, that the water at these profound depths lay dead and motionless, that no living thing could possibly exist at the bottom, and that no change went on there save the slow accumulation of fine mud, carried by ocean currents from far-distant land.

10. But much of this uncertainty has now been dispelled. Expeditions to measure the depth and explore

the bottom of the sea have been sent to all parts of the world, and have collected such a mass of information that we now know more about the depths of the ocean than about some parts of the continents. In particular, the ship *Challenger* was, in 1872, sent round the world by the British Government to ascertain the depths, temperature, and other features of the great ocean basins. Similar expeditions have since been organised by the United States and Germany. Three chief instruments have been used in these researches: the Sounding-line, the Dredge, and the Thermometer.

11. The Sounding-line consists essentially of a rope or wire with a weight at the end, so arranged that, when dropped from the side of a vessel to the sea-bottom, the amount of line which runs out from the reel is accurately registered, and at once indicates the depth. The weight may have its bottom covered with lard or other soft material, which, when it strikes against the bottom, will bring up some of the mud, sand, gravel, or other indication of the nature of the ground below. The Dredge consists of an iron frame with a strong net or bag attached to it; the object of this instrument being to bring up a larger quantity of the materials that form the sea-floor, as well as any of the living or dead plants and animals to be found there. The Thermometer has been used to such good purpose that the temperature of the vast Atlantic and Pacific Oceans has been determined at many points, from the surface down to the deepest parts of the bottom. By means of these instruments of research the first beginnings have been made towards a knowledge of the geography of the ocean-basins. The general results, so far as at present known, are expressed in Plate VII., which shows instructively how these basins are broken up by submarine ridges, and how connected the continental areas are in spite of the inland seas which divide them.

12. At present the tract of sea which has been most

thoroughly explored is the Atlantic Ocean. As shown upon Plate VII., this Ocean runs as a long and winding belt of water between the New World and the Old. Towards the north, as shown by recent soundings, it is closed in by a submarine ridge, which, extending from the north-west of Scotland through the Faroe Islands and Iceland to Greenland, separates it from the Arctic basin. Southward from the equator, the Atlantic Ocean gradually widens out and merges into the wide Antarctic Ocean. It has been estimated that the whole area of the Atlantic Ocean amounts to 35,165,000 square miles, which is nearly a fifth part of the entire surface of the globe.

13. Stretching from the Arctic to the Antarctic water, the Atlantic Ocean crosses the various zones of temperature which girdle the globe. Its central parts lave the shores of equatorial America and Africa. Beyond these, it reaches the temperate climates of North America and Europe. At the southern end it enters the icy Polar regions.

14. Again, owing to the arrangement of the continents which bound it on the west and east, the Atlantic receives a far larger river-drainage than any other ocean. The map shows that down the whole length of America all the large rivers flow eastward, and therefore fall into the Atlantic—the St. Lawrence, Mississippi, Orinoco, Amazon, and La Plata. In Europe and Africa, if we include the rivers which enter the Mediterranean and Black Sea, by far the largest proportion of the drainage finds its way into this ocean by such important rivers as the Rhine, Rhone, Danube, Dnieper, Nile, Niger, and Congo. The Atlantic basin is thus the great reservoir into which the largest rivers of the globe discharge their waters.

15. That the floor of the Atlantic cannot be a vast plain like its surface is shown by the islands which here and there rise even from the middle of the ocean.

From the numerous soundings which have been taken throughout its entire length and breadth, its form can be laid down with considerable accuracy upon a map. The wider parts of the Atlantic have a depth of from 2000 to 3000 fathoms, or from about 2 to $3\frac{1}{2}$ miles. They form a vast undulating plain, crossed by a ridge, of which the Azores form the highest elevation. This ridge may be regarded as starting from the western edge of the great plateau on which Britain stands. Passing southwards by the Azores it forms what is known as the Dolphin Rise, which, at its southern end, about latitude 30° N., ascends to within 400 fathoms from the surface. The ridge then strikes south-westward at a depth of less than 2000 fathoms to the coast of Guiana, whence it turns south-eastward across the ocean, coming to the surface under the equator in the lonely St. Paul's Rock, and turning southward as a long ridge from which the volcanic islet of Ascension and Tristan d'Acunha rise.

16. To the west of the British Isles for 230 miles, the slope of the Atlantic bottom is very gentle, being only six feet in the mile. But beyond that distance the ground descends more rapidly, for in the next 20 miles there is a fall of 9000 feet, or a descent of 450 feet in the mile, down to the level of the great submarine plain, which stretches for hundreds of miles to the west, with little variety of surface. This plain ascends slowly towards the north till it forms the great plateau which, culminating in the Faroe Islands and Iceland, separates the deeper water of the Atlantic from that of the Arctic Ocean. The Newfoundland banks prolong the North American continental mass far into the ocean. The Florida peninsula and West Indian Islands separate the deep Atlantic waters from the basins of the Mexican Gulf and Caribbean Sea, which are obviously submerged mediterranean seas (Art. 20).

17. In the deeper abysses of the Atlantic lying on both sides of the central ridge, soundings of between 3000

and 4000 fathoms have been found. About 100 miles north from the Island of St. Thomas the *Challenger* obtained a sounding of 3875 fathoms, or rather less than 4½ miles, in what appears to be a vast hollow running north-eastwards from the end of the ridge on which the West Indian Islands rise. This is the greatest depth yet found in the Atlantic Ocean.

18. Recent soundings in the Pacific Ocean have revealed the general contour of its bottom, of which indications are supplied by the distribution of the islands. In the South Pacific a submerged plateau extends westwards from Chili and Patagonia, and beyond it, to the west, a group of submarine peaks and ridges rises into scattered islets which culminate in Tahiti (7000 feet). New Zealand, New Caledonia, the Feejee Islands, and other groups, are based on a vast submarine platform between 1000 and 2000 fathoms below the surface, which, merging into that from which Australia rises, is prolonged towards the north-west through New Guinea, Borneo, and the Malay Archipelago into the great Asiatic area. An irregular winding ridge, supporting the Caroline, Marianne, and Bonin Islands, appears also to connect this platform with Japan. The oceanic islands in the North Pacific, including the Marshall, Gilbert, and Sandwich groups rise from similar submarine banks and ridges, reaching their highest point in the volcanic peaks of Hawaii (13,760 feet). It is worthy of notice that while the large islands on the prolongation of the Asiatic and Australian plateau (New Caledonia, New Zealand, and others) are composed mainly of non-volcanic rocks such as those of which the continents chiefly consist, the scattered oceanic islands, where they present any other material than coral-rock, reveal a volcanic origin. They have probably been originated by the piling up of volcanic rocks from submarine eruptions.

19. It is likewise deserving of remark that according to recent soundings the floor of the Pacific Ocean, like

that of the Atlantic, has an average depth of between 2000 and 3000 fathoms, but sinks here and there into deep basins, even among the plateaux and islands. The deepest abysses have been found between the Caroline Archipelago and the Aleutian Islands. In particular, immediately to the east of Japan, lies a long trough, of which the bottom descends to between 4000 and 5000 fathoms. There the *Tuscarora* took the deepest sounding yet obtained in any part of the globe,—4655 fathoms. In the Caroline Archipelago the *Challenger* found the depth to be 4475 fathoms.

20. The map of the world shows that although the great water envelope of our planet may, for the sake of convenience, be parcelled out into separate oceans, these are all united into one vast continuous sheet of water. Here and there, however, owing to the way in which the land has been ridged up, portions of the water have been almost separated from the main mass. These are called *mediterranean*, the best example being that which has long been known as the Mediterranean Sea. The Black and Baltic Seas in Europe, Hudson's Bay and the Gulf of Mexico in North America, the Red Sea, Persian Gulf, and Seas of Japan and Okhotsk in Asia, are other illustrations. It will be observed from Plate VII. that such inland seas are comparatively shallow, and really belong not to the oceanic but to the continental areas of the earth's surface. Though their sites are now occupied by the sea, they may once have been land, and might be raised into land again without greatly disturbing the present order of things.

21. But sometimes the uprise of the land has taken place in such a way as to cut off completely some outlying parts of the oceans. These are termed *inland seas*, of which the Caspian Sea and Sea of Aral are the most important instances. The former sea covers a larger space than the British Isles. Its surface is about 85 feet below sea-level, and its greatest depth amounts to nearly 3000

feet. Its waters are tenanted by seals and other animals that elsewhere inhabit the ocean. That a much larger area in that region was once submerged is shown by the fact that in the tracts of land which now inclose the Caspian and Sea of Aral and separate them from the Black Sea on the one hand, and from the Arctic Ocean on the other, beds of dead sea-shells are found. The sea has been excluded by the rise of its bottom into land. Reference will be made in Lesson XXIII. to the proofs that the land along the coast of Siberia has in comparatively recent times been raised out of the sea. It has even been conjectured that the Arctic Ocean formerly extended in a long arm between Europe and Asia as far as the hill-range which is now cut through by the narrow channel of the Bosphorus, but did not communicate with the present Mediterranean Sea, and that by the rise of the land towards the north all that part of this vast inlet lying to the south of the parallel of 50° or 52° N. was cut off from the main ocean. The present abundant salt lakes and marshes, as well as the two large basins of the Caspian and Aral, have been regarded as the mere shrunk remnants of this old Mediterranean Sea. The Black Sea has been separated from the waters of the Caspian region, and the intervening ridge between it and the Mediterranean Sea has been cut through, so that the Black Sea now communicates through the Bosphorus and Sea of Marmora with the Mediterranean. There seems also to be less rain or more evaporation now than formerly in the region of the Caspian and Aral, so that these sheets of water are still further shrinking.

LESSON XIII.—*The Saltness of the Sea.*

1. The aqueous vapour of the atmosphere is in great part derived from the evaporation of the surface of the

sea, and as the sea does not become sensibly lower in level, notwithstanding the enormous volume of moisture that is annually driven off from it into the air, it must receive back again from the rivers that descend to its shores, and from the rain that falls upon its surface, as much water as it supplies. There is thus a ceaseless coming and going of water between the sea, the air, and the land.

2. When the water, however, is examined in different parts of its passage, it shows great varieties. Though rain, being water condensed from natural distillation, is nearly pure, it sometimes has a distinct taste, and when evaporated leaves behind various salts and particles of dust, which it has taken out of the air (Lesson VI.). The water of springs and rivers gives a still larger proportion of substances after evaporation, though usually quite fresh and tasteless. But sea-water is always strongly salt to the taste, and this is true of the sea in all quarters of the globe.

3. Another distinction between the water of the air or of springs and rivers and that of the sea is shown in their different weights. A bottle filled with sea-water weighs more than when filled with fresh-water. Perfectly pure water being taken as the unit, sea-water has an average relative weight or specific gravity of about 1·026; in other words, if a certain quantity of perfectly pure water weighs 1000, the same quantity of sea-water will weigh 1026. The water of the different oceans varies slightly in density. That of the Atlantic, for example, varies more and is generally heavier than that of the Pacific. In the path of the trade-winds in the North Atlantic, where evaporation must be comparatively rapid, some of the heaviest and of course saltest sea-water occurs, its specific gravity being 1·02781. In the Antarctic Ocean, on the other hand, among the broken ice, the specific gravity sinks to 1·02418.[1]

[1] Buchanan, *Proc. Royal Society* (1876), vol. xxiv.

THE SALTNESS OF THE SEA.

A striking proof of the greater lightness of fresh water may sometimes be seen after heavy rain has fallen in still weather upon the sea. The rain floats on the surface of the sea, and so does the water brought down by streams from the land. Fresh drinkable water may then be taken from the surface of the salt-water which lies below, until, unless more quickly mingled by wind and waves, the fresh-water slowly diffuses itself downward.

4. What is it that makes the sea-water heavy and salt? To answer this question let a small quantity of

FIG. 13.—What is seen when a drop of concentrated sea-water is evaporated under a microscope.

sea-water be evaporated till only a little of it remains, which tastes extremely salt and bitter. Stirring this remaining liquid, so as to mix it completely, put a drop or so upon a piece of thin glass under a microscope, and watch what takes place in this simple and beautiful experiment. The drop continues to pass off into vapour, but the substances which gave the water its salt and bitter taste cannot do so. They remain behind, and as the water leaves them they are seen shooting into symmetrical crystals. Wonderfully interesting is it to watch how the little particles rush to each other, and unite to form those exactly proportioned outlines. First there appear some oblong, pointed forms (gypsum, or sulphate of lime), which probably began to form before the drop was taken out of the concentrated salt-water; then, as the drop rapidly evaporates, a crowd of little squares marshal themselves all round the edges, and then become fixed and motionless. The water has been driven off into vapour, and what remains on the glass is the salts that give the sea-water its characteristic taste and weight. Those who live far from the sea may see a part of this experiment by taking a drop of water into which as much common salt has been put as it will dissolve, and placing it under the microscope. The square crystals will be seen growing round each other with great vigour and in perfect symmetry.

5. While the different forms of crystals indicate the different saline materials present in the sea-water, each substance which crystallises keeping its own form of crystals, the relative abundance of each form shows in a general way the proportions in which the substances occur in the water. We see, for example, that by much the most abundant are the square or cubical crystals. These belong to common salt, or chloride of sodium. They make their appearance after the gypsum has crystallised, whence it is clear that they can remain longer dissolved in the water. They are therefore said to be more soluble. Now while the dry film left by the drop is still below the microscope, breathe gently on it, and watch what takes place. The moisture of your breath dissolves the crystals one by one, until, if you have supplied moisture enough, they wholly disappear in the little drop of water which now replaces them. The cubical crystals of salt vanish first. Should there be much vapour in the air, you may not be able to keep the crystals from attracting it and dissolving, unless you heat the glass before a fire. This tendency to attract moisture and pass into solution is called **diliquescence**.

6. When sea-water is analysed it is found to contain

about three and a half parts by weight of salts in every hundred parts of water. As indicated by the drop evaporated in the experiment, common salt forms by much the largest proportion, at least three-fourths, of the whole amount of salts. The other saline substances are chlorides of magnesium and potassium, sulphates of lime (gypsum) and magnesia (Epsom salts), and some others in still smaller quantity. The proportions do not greatly differ throughout the whole extent of the ocean. As the sea receives the rivers that descend from the land, its waters must contain in solution the various substances carried in river water. There is perhaps no kind of mineral matter in the outer portion of the earth which does not occur dissolved in sea-water, though the proportions of them are usually so minute as to escape detection by the ordinary process of analysis.

7. The next question that arises is, Whence does the sea get its salt? This is not quite so easily answered. We cannot tell what the earliest condition of the sea may have been. Probably its waters have always been salt, ever since they condensed out of the first atmosphere of gas and vapour that enveloped the earth. The saline vapours, which were no doubt diffused abundantly through that atmosphere, would be carried down in solution into the primeval ocean. But even if the sea had been quite fresh at the beginning, it would have been perceptibly salt now. In Lesson XXVII. an account will be given of the way in which the fresh-waters that flow off the land invariably carry saline substances to the sea. As this process continues without ceasing, and in every quarter of the globe where land exists, no one can fail to understand how vast must be the yearly tribute of dissolved mineral matter received by the sea from the land. If the methods of analysis were delicate enough, it would be possible to detect in sea-water traces of every substance dissolved in the waters of the springs, rivers, and lakes of the land. Some ingredients, for example,

which are present in too minute proportion to be detected by chemical analysis in a small quantity of sea-water, are absorbed by marine plants and animals, and found in their ashes or skeletons. The copper bottom of a ship which has been sailing for months gathers out some of the dissolved arsenic or other metals which are diffused through the waters of the sea, and these substances are likewise found in the crust which forms inside the boilers of ocean-steamers where salt-water is used.

8. Among the ingredients present in sea-water in such minute proportions as almost to escape notice, two are of such importance that their names should be noted— **carbonate of lime and silica.** The former of these is the material of which the hard parts of shells, corals, sea-urchins, star-fishes, and other familiar sea-creatures are mainly built up. When embodied by these animals as part of their skeletons, it becomes the white chalk-like substance which we recognise as the material composing the common shells of the sea-shore. Silica is secreted by lowly forms of plant and animal life in the sea. The hard compact stone called flint is a form of ancient sea-grown silica.

9. Sea-water likewise contains **air.** That this must be the case is evident from what takes place in windy weather at sea. The surface of the water is tossed up into waves, and these are blown into spray. The sea becomes a sheet of white foam, which is a mixture of air and water. Some of the air remains dissolved in the water after the waves have disappeared, and is slowly diffused even to the depths of the ocean. One important function of wind is thus to aërate the water of the sea. The quantity of air present in sea-water varies from time to time; sometimes it has been found to amount to no more than one-hundredth of the volume of water, at other times to as much as one-twentieth.

10. These may seem small proportions, yet we cannot

but look upon them with interest, for it is this included air which enables the animals of the ocean to breathe. They inhale it, and exhale carbonic acid. Hence that gas, besides the proportion of it in the dissolved air, is likewise present in sea-water. In the higher parts, which come under the influence of wind-waves, carbonic acid is partly removed, and fresh supplies of air are furnished in its place.

11. One other substance in sea-water deserves notice here, called **organic matter**, or **protoplasm**. This is the material which serves the simpler forms of marine creatures as food. It is derived from the substance of dead plants and animals, which live in the sea, and of those which are carried by rivers from the land.

12. It appears, then, that the sea is salt, partly probably from the vapours of the original atmosphere from which it condensed, partly because many different salts are continually carried into it by brooks and rivers; that it contains in solution more or less of all terrestrial substances which water will keep dissolved; that it is aërated by the waves, the air that is diffused through its mass serving for the respiration of marine animals; and that from the dead bodies of animals and the remains of plants, both on land and sea, the fundamental substance or protoplasm of which plant and animal forms are built is diffused through the whole mass of the ocean.

LESSON XIV.—*The Depths of the Sea.*

1. From what has been said in Lesson XII. it is plain that the bottom of the sea must be uneven and undulating, like many parts of the surface of the land; stretching out in some parts into vast plains, rising here and there into broad and long ridges, shooting up into vast slopes, like that of the Atlantic floor to the west of Britain, sinking into deep hollows, which descend far below its general

level, as lake-bottoms do below the general level of the land, and traversed by mountain ranges, of which the oceanic islands are only the unsubmerged peaks. But what is the nature of the surface of the sea-floor? Is it solid rock, or soft mud, or barren sand,—or are the submarine plains covered with an oceanic vegetation, as the plains of the land are with grass and herbage?

2. Standing by the margin of the sea we observe that the water breaks upon sand, gravel, mud, or shells, and that these materials pass down beneath it. If the shore is rocky, pools of the salt-water may be noticed, from which some idea may be formed of the nature of the bottom of at least the shallower parts of the sea. Each of these pools forms as it were a miniature sea. Its sides, hung with tufts of delicate sea-weeds, are bright with clusters of sea-anemones, while many a limpet and periwinkle rests fixed to the stone, or creeps cautiously over its surface. The bottom of the water abounds in shady groves of algae, through which many tiny forms of marine creatures dart or crawl. As we look into one pool after another, we find them all to be more or less full of plant and animal life.

3. Turning from these shore pools to the edge of the sea itself when the tide is low, we mark that the ledges of rock or sheets of sand and shingle support a thick growth of coarse dark green or brown tangles and sea-wrack, among which, if the water is still enough, tiny crabs, sea-urchins, jelly-fish, and other bright-coloured marine animals may be seen. If the water is examined from a boat, this forest-belt of large dark sea-weed is found not to extend to a greater depth than a few fathoms. Beyond it the bottom, whether rocky, sandy, or muddy, can be seen through the clear water, or may be examined by means of the dredge. Delicate scarlet sea-weeds with corallines and deeper-water shells inhabit these tracts. The sea-weed belt which fringes the land has an average breadth of about a mile. Beyond it, as

we gradually get into deeper water, the common plants and animals of the shore are found one by one to disappear, and other kinds to take their place. The dredge may be dragged along some parts of the sea-floor and bring up only sand or mud, while at a short distance off it may come up full of many and varied forms of marine life, thus showing that there must be bare tracts of sand, mud, or stone on the sea-floor, and other patches where plants and animals are crowded together.

4. Down to a depth of 100 fathoms, the waters of the ocean round the margin of the land abound in plants and animals. This upper and marginal belt supports indeed by far the most abundant and varied marine population. Down to depths of 200 or 300 fathoms there is still a plentiful assemblage of animals. But as the depth and distance from land increase, not only do the types of the shallower waters diminish, but other forms make their appearance which are universally diffused over the deeper parts of the ocean. Until recently, naturalists supposed, that as life becomes less prolific in the deeper water, the great abysses of the sea must be barren of life. They even fixed the limit at 300 fathoms, below which, as light ceases to penetrate farther, they thought the waters of the ocean must be lifeless.

5. But within the last few years, researches, carried on across the Atlantic and other oceans, have shown that no such limit really exists, that, on the contrary, an assemblage embracing representatives of all the great types of animal life exists on the sea-floor at depths of more than two miles, where total darkness must prevail, and where the temperature is very low. The species of animals obtained from these depths, however, differ from those of the shallower tracts. They present no great variety of form, nor generally a great abundance of individuals. Plant-life is much more restricted in its range. Sea-weeds abound from high-water mark down to about 15 fathoms; they become rare beyond a depth of

50 fathoms, and are not found at all in water deeper than 200 fathoms, although some other kinds of marine plants occur at still greater depths. But from the bottom of the profounder abysses plants seem to be wholly absent.

6. In such comparatively shallow water as that which surrounds the British Islands, where the action of sea-currents extends to the bottom, banks of sand, with patches of broken shells and gravel, and wide flats of mud, cover much of the sea-floor. Some of these sand-banks, where oysters, clams, and other shell-fish live, furnish good feeding-ground for cod, haddock, turbot, and other familiar fishes, and are therefore well-known to fishermen. They seem to retain their place and form for many years. But in the shallower parts of these seas they are apt to be altered by storms. Some of them come so near the surface that vessels are often driven upon them and wrecked.

7. The mud, sand, and gravel, strewn over these tracts of the sea-bottom, all consist of water-worn particles of rock. If we could trace back the history of these deposits, we should find that each grain of sand and each pebble of gravel once formed part of the solid rocks of the land. It will be pointed out in Lesson XVIII. that these materials are not produced on the sea-bottom, but are brought thither from the land. It is along the margin of the land, where waves can act upon the rocks, that the sea does nearly all its erosive work. Every stream, too, is busy carrying and pushing mud, sand, and gravel to the sea. Thus, whether the materials have come from the surf-beaten shore, or from the hills and valleys of the land, they are strewn over those tracts of the sea-bottom which lie nearest to land. The sea is, indeed, the vast receptacle into which all the materials derived from the crumbling of the surface of the land must ultimately find their way. Wherever there is movement of the bottom water, sand, and even fine gravel, may be pushed along. But, throughout most

of the great sea-basins, the water that lies upon the bottom, though not quite motionless, has too little motion to be able to grind down rocks into sand and mud, and is probably too sluggish in its flow even to disturb the fine sediment which must be the only kind of deposit that can be slowly accumulated there.

8. Much light has been thrown upon the condition of the deep-sea bottom by the soundings of the *Challenger* expedition. In comparatively shallow water near land, whether islands or continents, variously coloured deposits of mud and sand occur, evidently in great part formed from the finer sediment which has been worn away from the surface of the land and carried out to sea. These deposits of land-derived sediment may extend in exceptional cases to 300 miles from the continents, though usually confined to much narrower limits. But away from land, in the wide and deep depressions of the ocean basins, other characteristic deposits are laid down. The most universal of these are fine red and gray clays, in which fragments of the volcanic rock called *pumice*, may be recognised. Remains of minute organisms (*Globigerina* and *Radiolaria*) occur in them, sometimes abundantly. Most of the bottom of the Pacific and Atlantic basins, at depths of more than 2000 fathoms, as far as yet explored, is covered with these clays.

9. There seems little reason to doubt that these fine clays are derived from the decomposition of volcanic detritus slowly drifted across the ocean. This detritus may come mainly from submarine eruptions; but part of it is no doubt supplied by the pumice which is sometimes discharged from volcanic islands in prodigious quantities, and which, being light and porous, may float away to vast distances before becoming water-logged and sinking to the bottom. The extreme slowness with which the clays gather on the ocean-floor is strikingly shown by the fine iron dust of falling stars which has been recognised in an appreciable quantity in the clays brought up in the

Challenger soundings, and by the finding of 600 sharks' teeth, 100 ear-bones of whales, and other bones of whales, in a single haul of the dredge. These objects have accumulated with almost inconceivable slowness on the sea-bottom, yet have not been covered up in the still more slowly gathering clays.

10. Other wide-spread oceanic formations are derived from the accumulation of the remains of minute plants or animals. In the Southern Ocean, towards the icy Antarctic land, a pale, straw-coloured deposit has been dredged up from depths of nearly 2000 fathoms.

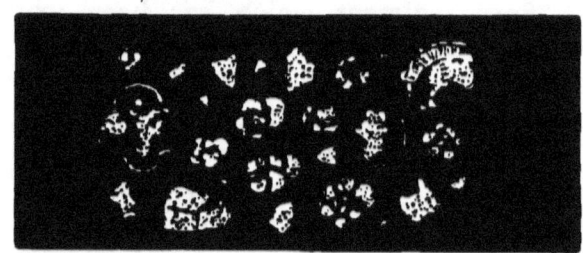

Fig. 14.—Some of the Ooze of the Atlantic floor, magnified.

It consists chiefly of the remains of minute siliceous plants called *diatoms*. Another kind of siliceous deposit, formed from the minute animal forms called *Radiolaria*, covers a considerable space in the intertropical regions of the Pacific. The best known of these deep-sea formations is the **ooze** which was brought to light when the earliest soundings were made for the telegraphic cable between Britain and America. It covers the floor of the Atlantic for thousands of square miles as a fine, soft, chalky deposit. A sounding-lead dropped upon the bottom sinks deeply into it. When the dredge goes down, it plunges into this light, fleecy mud, and comes up full of it. The upper layer is a yellowish, creamy, somewhat sticky substance, full of entire and broken shells of such minute animal organisms (*Globigerina*) as are shown in Fig. 14, with fragments of sponges and other forms. Underneath this surface layer the ooze shows the shells in a more crumbling state, and no doubt in the deeper parts of the deposit, which the dredge cannot reach, the substance may consist only of a kind of fine chalk, formed almost entirely of the mouldered remains of these lowly forms of life. Living sponges, star-fishes, crinoids, and various shells have been brought up by the dredge with the ooze. It would seem that while this widely extended deposit may contain the remains of such forms of life as inhabit deep water, it is derived mainly from the *Globigerina*, which lives in immense numbers in the surface waters, and which, when dead, sinks through the depths to the bottom.

LESSON XV.—*The Temperature of the Sea.*

1. Since the temperature of the air and of the land varies in different latitudes, we may expect to find that of the sea to show corresponding variation. There are indeed climates in the sea as well as on the land. The temperature of the surface water of the sea varies with the seasons, and with the direction of the great ocean currents. Thus, on the west side of Britain, the superficial sea-temperature is about 49° Fahr., and over a great part of the North Atlantic it ranges between 44° and 54°. Further south, in the equatorial part of that ocean, it rises to from 80° to 83·5°. In the North Pacific Ocean the mean surface temperature is about 70°, and in the South Pacific about 67·7°. In Polar seas, away from the influence of warm currents, the water does not rise much above the temperature of melting ice. On the other hand, in land-locked parts of the ocean within the tropics, where the water is confined, and exposed to great solar heat, the temperature sometimes rises to that of a warm bath. Thus, in the Red Sea, it has been noted at 94°, and in different parts of the Indian Ocean at from 88° to 91°.

2. The importance of the temperature of the sea is seen in the influence which it exerts upon the climate of the maritime tracts of land. The temperature of these regions depends greatly upon that of the winds which come from the sea. The air resting upon the ocean surface acquires the temperature of that surface, and carries it over the land. Save for this convection, the sea would have comparatively little influence upon the climates of the land. But by warming or cooling the air over tracts thousands of square miles in extent, it becomes the great distributor and regulator of temperature. This has been partly explained in Lesson XI., and will be further treated of in Lesson XVIII.

3. The surface temperature of the sea varies with the latitude and with the season of the year. Over the wider oceans the isothermal lines for the surface-water do not greatly vary from the direction of the parallels of latitude; but in the North Atlantic they depart widely from that direction, for the reasons already explained (Lesson IX. Art. 13). The surface-waters of the sea in the northern hemisphere are coldest in February and warmest in August; in the southern hemisphere they are coldest in August and warmest in February. Moreover, the surface temperature varies at each place during every day, perhaps to the amount of rather less than one degree Fahr., the lowest temperature being reached in the morning and evening, and the highest between two and four o'clock in the afternoon. Recent observations of the *Challenger* expedition, discussed by Mr. Buchan, indicate that the daily oscillation of the temperature of the air over the sea is $3\cdot21°$ Fahr., or nearly four times greater than that of the water on which it lies.

4. It is evident, however, that the readings of a thermometer immersed merely in the surface oceanic water give no more indication of the real temperature of the great mass of the sea than if we tried to estimate the temperature of the whole atmosphere merely

from observations with the thermometer at the sea-level. By taking observations at intervals up to the mountain-tops, the gradual lowering of the temperature referred to in Lesson IX. Article 20, has been determined. In like manner, by means of a specially arranged kind of thermometer, the temperature of the sea at any depth can be ascertained. The instrument is protected in such a way as to be unaffected by the pressure of the water. It is let down with a cord, allowed to remain long enough at each depth to acquire the temperature of the surrounding water, and then pulled up.

5. In this way the temperature of the Atlantic and Pacific Oceans has been determined at intervals from the surface down to the deepest depressions of the bottom. Temperature-soundings have likewise been made in the Indian Ocean. As the result of such observations all over the world, it has been discovered that in general the surface-water is the warmest part of the sea, and that the water gets colder towards the bottom. In the open sea, indeed, communicating freely with the Polar regions, the deep bottom-water has everywhere been found to have very nearly the same temperature as that of the cold Polar Seas. Sometimes the thermometer has come up with readings as low as $29 \cdot 5°$; that is, two and a half degrees colder than the freezing-point of fresh water.

6. Soundings taken by the *Challenger* showed everywhere a progressive lowering of the temperature from the surface downwards, with variations in the rate of decrease. In the North Atlantic, water having a temperature above $40°$ forms the upper layer of the ocean to a depth of from 750 to 1000 fathoms. Under that warmer sheet the temperature slowly sinks, until in the deep abysses, more than 3000 fathoms down, it reaches $34 \cdot 4°$. Hence the main mass of this ocean consists of chilly water. In the South Atlantic the warm upper layer is only about 300 fathoms thick. Below it the temperature of the water falls from $40°$ to $32 \cdot 9°$, so that

the body of chill water is colder and comes nearer to the surface than in the North Atlantic. In the equatorial part of the ocean, while the surface temperature is 76°-80°, the cold water under 40° is found at a depth of only 300 fathoms. The bottom layer of water is at less than 35° for a thickness of 600 fathoms, while some of its deeper cavities have the chilly temperature of 32·4°.

7. In the North Pacific Ocean a general bottom temperature below 35° Fahr. has been observed. Sometimes the body of water at this low temperature exceeds 2000 fathoms in depth. In the South Pacific Ocean, beyond Cape Otway, Australia, the bottom temperature sinks to 32·5°, while farther south (Lat. 53° 55′ S.), the body of icy water below 35° comes within less than 100 fathoms from the surface. Below a depth of 1000 fathoms the water is as cold as ice (32°) down to the bottom, where the thermometer marked 31° at a depth of 1950 fathoms.

8. The bottom water may be regarded as remaining constantly at an almost icy temperature, with a remarkably small variation of only about 7°, or from about 31° to 38° Fahr. This low temperature of all but the upper parts of the sea shows that as the deep water could not be cold unless it came from cold latitudes, there must be a continual transference of water from the cold Polar tracts towards the equator. The heavy cold water sinks down, and creeps on below the upper warmer layers, which move towards the poles to supply its place. When the cold water reaches the equatorial regions, it must rise towards the surface, where, being gradually warmed, it moves away again as a surface layer towards the poles. Hence soundings with the thermometer in the oceans not only reveal the various temperatures of the sea, but bring before us evidence of vast slow movements, by which the oceanic waters are constantly mingled and kept from stagnation.

9. In the North Atlantic, between Scotland and the

Faroe Islands, below a uniform surface temperature of 50°-53° Fahr., two parallel belts of water have been observed lying side by side and having nearly the same depth. In the southern of these the bottom temperature was found to be 42·7°, in the northern 29·6°. Two very different sea-climates were thus ascertained to lie close together, one with the mild character of the latitude, the other with the chilly temperature of the icy Arctic sea. It was at first supposed that two parallel currents here impinged against each other, one coming from the icy north and the other from the warmer Atlantic. But recent researches have shown that the warm and cold areas are separated from each other by a submarine ridge which completely prevents any flow of polar water into that part of the North Atlantic. So that the cold water with its characteristic forms of plant and animal life is banked up against the north side of the ridge, while the warmer Atlantic water, with its distinctive organisms, lies at corresponding depths on the south side.

LESSON XVI.—*The Ice of the Sea.*

1. When sailing across the North Atlantic between ports in Europe and in the United States, or when traversing the South Pacific to round the southern promontory of the American or African continent, vessels occasionally encounter floating masses of solid ice, termed **Icebergs**. These are of all sizes up to huge mountain-like islands rising several hundred feet out of the water. Ice being lighter than water, an iceberg projects above the sea-level, yet has about eight times more of its bulk under water than above. Hence, when the mean height of one of these floating masses is 300 feet above the surface of the sea, its bottom must be some 2400 feet below. The bergs vary infinitely in shape as well as in size. Sometimes, as in the Antarctic Ocean, they take

the form of vast square blocks with vertical walls; sometimes they bristle into peaks and pinnacles, with deep cliffs and gullies between. At a distance they look like snow-covered islands. Seen closer, they gleam with all the intensity of colour — white, green, and blue — so characteristic of the ice of glaciers upon the land. For the most part, nothing but ice in different forms is seen upon their surface. Now and then a block of stone or

FIG. 15.—Iceberg at Sea.

some dark earthy rubbish may be noticed. As they drift onwards into warmer latitudes they melt away both under water and above. Cascades of water tumble down their thawing slopes and fall into the waves below. It often happens that, as melting advances under water, the centre of gravity of an iceberg is altered, so that the mass shifts its position, or, becoming top-heavy, turns completely over.

2. Such are the drifting icebergs which in summer cross the navigation track across the Atlantic between

Newfoundland and Britain. They form a serious source of danger to vessels, for they may be encountered at any hour of the voyage. Many a ship has struck against one in the dark and gone to the bottom. They chill the air around them, and thus often give rise to dense fogs. A sudden fall of temperature is regarded by sailors as an indication of their approach to an iceberg, and a warning to be on the outlook.

3. But icebergs are not formed in the open sea, where they float until they are entirely melted. To reach their birthplace we must travel to the cold regions which surround each pole. In North Greenland, for example, the country is covered with ice, which, creeping slowly down the slopes, gathers in the valleys into wide and deep masses that not only come down to the sea-level, but actually push their way out to sea in vast walls of ice, rising 300 or 400 feet above the water, and stretching sometimes for 60 miles along the coast. From time to time portions of these great tongues of ice break off and float out into the open sea. These are icebergs. They consist, therefore, not of frozen sea-water, but of true land-ice. Each berg which may be met with, floating and melting far away in mid-ocean, had its origin among the snows of the Polar lands. (Lesson XXVIII.)

4. Since so large a proportion of their bulk is under water, we may expect to find that floating icebergs are comparatively little influenced by winds and waves, for these touch only their upper parts. The larger bergs move with the ocean-currents or drifts in which so much of their substance is immersed. Hence they may sometimes be seen moving steadily and even rapidly along, right in the face of a strong gale.

5. So long as the icebergs sail over deep water, they move freely about, as the currents or winds may drive them. But when they get into water shallow enough to allow their bottoms to grate along the sea-floor, they tear up the mud or sand there, until they are at last stranded.

Fig. 16.—The Birthplace of the Arctic Icebergs.

The coast of Labrador is often fringed with such grounded bergs, some so small as to be driven to the beach, others so large as to run aground while still a good way from the shore. Chill fogs consequently hang over that desolate region all through the summer.

6. Besides icebergs, other kinds of ice are found on the sea in high latitudes, derived from the freezing of the water of the sea itself. As voyagers advance into the narrow seas about the pole, they encounter great sheets of ice, which, at first in irregular, scattered fragments, become larger in size and more abundant until at last the whole expanse of the sea, as far as the eye can reach, is covered with ice. This is known as **Floe-ice** or **Field-ice**, and is due to the freezing of the surface-water of the sea during the intense frost of Polar winters. When the sea-water freezes, the salt is left behind by the ice, which, except for the little salt vesicles entangled in it, is fresh, and, when thawed, quite drinkable. In early summer, the ice-field, or sheet of ice, which, stretching outwards from the land, may be regarded as continuous for hundreds of miles during winter, begins to break up and float away. If the sea remained motionless it would, by the end of winter, be covered with a sheet of ice about eight feet thick. But owing to the pressure arising partly from the movements of tides and currents in the water, partly from changes of temperature and atmospheric pressure, including winds and storms, the ice-sheet is liable to continual disruption. Cracks are formed abundantly through it, and portions are squeezed up or broken into innumerable huge fragments, which are pushed over each other until the frozen sea becomes like a heap of ruins. These movements are accompanied with sharp reports, as of cannon, and with loud growling noises, as if some monsters of the deep were engaged below in fierce and angry warfare. Some notion of the appearance of this broken-up ice may be formed from Fig. 17. This represents a scene in the

memorable Arctic voyage of the Austrian vessel *Tegetthoff*, which, after almost incredible endurance, was eventually abandoned by the courageous crew, who, taking to

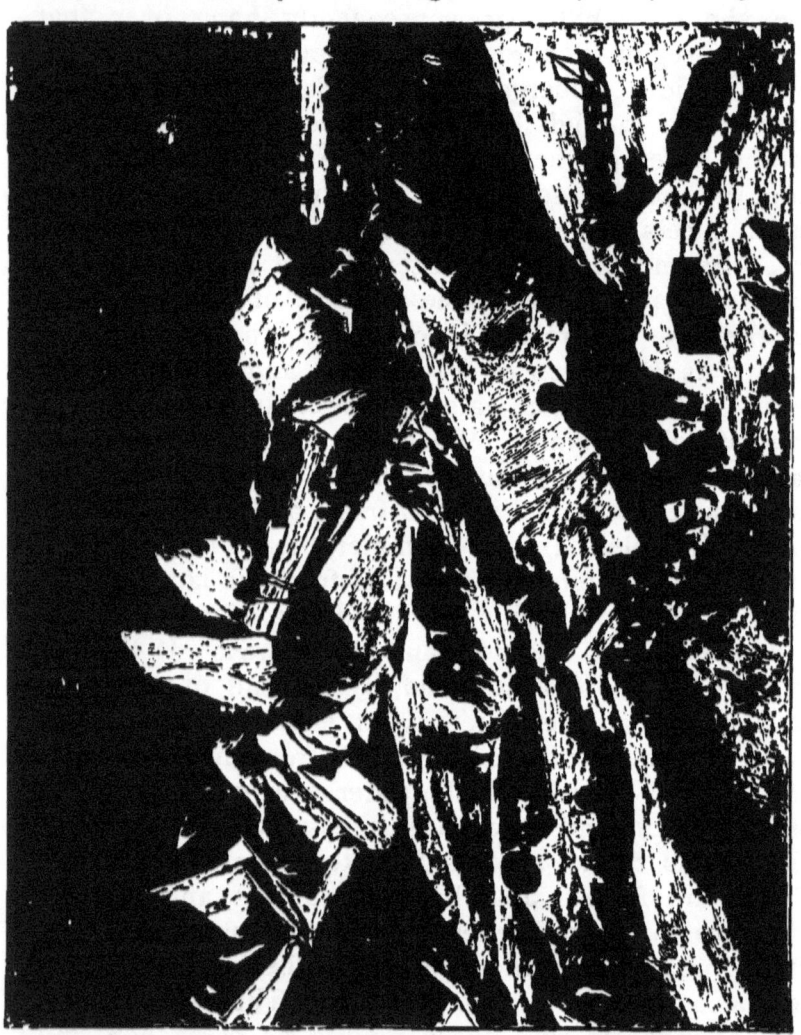

Fig. 17.—Scene among the disrupted ice of the frozen Arctic Sea.

boats or sledges drawn by dogs, pushed across the frozen sea and reached Novaya Zemlya.[1]

[1] For a graphic account of this and other Arctic voyages, see *New Lands Within the Arctic Circle*, by Julius Payer.

7. In consequence of the breaking up of the vast sheet of ice, water-channels or lanes are opened out between the separated fragments, and form passages, by which vessels make their way through the ice-pack. But as the great sheets are carried against each other in the general movement, it sometimes happens that a ship is caught between them, and pushed up on the floe, or crushed so effectually that it goes to the bottom as soon as the fragments of ice separate again.

8. Except where piled in heaps by pressure, which breaks up its surface as well as its outer edges, floe-ice occurs in level sheets, the surface of which rises but little above the level of the sea. It never rivals the height and grandeur of true icebergs, though it covers a much wider space of sea. Nor does it travel so far from the regions of its birth. When the ice-field breaks up in summer the portions next the land may remain there, and of these, indeed, some parts may continue unmelted for generations. But other parts farther from shore, and sometimes many hundreds of square miles in extent, are loosened and carried away by the sea-currents which drift from the pole. Vessels frozen into the ice-field have in this way been carried many hundreds of miles, until disengaged by the disruption and thawing of the ice. Some of the icebergs from Greenland, however, travel much farther south, and may be met with now and then even as far south as lat. 37°, that is, about the same parallel as Richmond in Virginia, and Cape St. Vincent in the south of Spain.

9. When the sea freezes along the margin of the land, as it does in a remarkable way in North Greenland, it forms a cake of ice which, rising with the tide, is frozen to the shore. By degrees a shelf of ice, called the Ice-foot, rising from twenty to thirty feet above the general level of the floe, and having a width of 120 feet or more, forms along the coast and clings to it all winter. Immense quantities of earth and stones, dislodged from the coast

cliffs by the severe frost of the Arctic winter, fall upon the ice-foot, so that its surface becomes in some places a waste of rubbish which completely covers and conceals the ice underneath. When the summer storms arise, this shore-terrace of ice is broken up, and large pieces of it, laden with the waste of the cliffs, are floated away out to sea, among the fleets of bergs and broken sheets of

Fig. 18.—The Ice-foot of Greenland.

floe-ice. Some portions are driven ashore again, others are caught and frozen fast into the floe-ice of next winter, while others succeed in escaping into the more open sea, where they gradually melt and tumble their load of earth and stones on the sea-bottom.

10. Floe-ice and the ice-foot are formed by the freezing of the surface of the sea. There is yet another way in which some of the ice of the sea takes its origin, viz., by the freezing of the water lying on the sea-bottom. This

is known as **Ground-ice.** It is probably formed only in enclosed and shallow seas and inlets, and is of little consequence compared with the thick and wide sheets of floe-ice. It is well known in the Baltic Sea. In still weather, before the surface of that sea is frozen, little thin cakes of ice may be observed floating about, sometimes with portions of sand or scattered pebbles imbedded in them. These are formed on the bottom, from which they break away and rise to the surface. They form a source of some danger, at least to small boats, for they sometimes appear suddenly in such numbers as to cover the sea, the surface of which is then apt to freeze too, so that the boats are in danger of being nipped between the detached ice-rafts, or of being beset and frozen into the united cake of ice. Sometimes large blocks of rock as well as masses of seaweed are borne away from the sea-floor and carried to the surface by the ascending sheets of ground-ice. In the rivers of cold countries, as for instance, in the St. Lawrence, similar ground-ice is formed in winter, sometimes round iron chains or anchors, which, if the enclosing ice is thick enough, may actually be lifted by it. Hence it is known as **Anchor-ice.**

LESSON XVII.—*The Movements of the Sea.*

1. The restlessness of its water is one of the features of the sea which first impress the onlooker. Even in a calm day, when not a leaf appears to be stirring on the earth, and the clouds seem perfectly motionless in the sky, the sea may be seen heaving, or curling into faint ripples, or rolling in broad undulations which break into foam when they reach the shore. On the other hand, in stormy weather the unrest of the air is fully equalled by the tumultuous uproar of the sea, when clouds of driving rain and salt sea-spray are swept along by the tempest, until it almost seems as if sea and sky were commingled.

2. We do not at first perceive that the movements of the sea are not all by any means so fitful and capricious as they appear to be, but that, on the contrary, they are regulated by easily understood laws, and can, to a great extent, be foretold and, if need be, provided against. Nor do we at once realise to what different causes these movements are due.

3. Let us suppose that, placed in some favourable position for observing the motions of the sea, we carefully register the facts that come under our notice. The margin of any of the great oceans would be suitable for the purpose. Such a coast-line as that of the west of Ireland, for example, forms an excellent field of observation, allowing opportunities of watching some of the most characteristic movements of the water of a great ocean basin.

4. The first point to be remarked is the obvious close dependence between the motion of the air and that of the water. As a rule, when the air is still the water is smooth, though on some days we should find that even with no wind, a good deal of agitation might be observable in the sea, large waves rolling heavily to shore. In such cases, however, we may often learn that a gale, blowing somewhere out at sea, has raised the waves, of which the final undulations now reach the land. Every gradation of ripple, wave, and billow may be traced on the water, answering doubtless to gentler or stronger movements of the air. In this respect the surface of the sea shows the same sensitiveness as that of any lake or pond on the land, but its movements are more marked, because it is so much larger, and allows the accumulated effects of the continued pressure of the wind to make themselves far more apparent.

5. But we could not long attend to the aspects of the sea without observing traces of other movements independent of the daily or hourly variations in the state of the atmosphere. On the Irish coast, which we have

taken by way of illustration, as on other parts of the seaboard of north-western Europe, scattered leaves and fruits are occasionally washed ashore from the wide ocean, which differ greatly from the vegetable productions of any European country. These are in reality West Indian plants. They have been drifted across the Atlantic, and hence they indicate the existence of some general drift or current which sets across that ocean from west to east or north-east. In like manner, bottles containing pieces of writing, which have been thrown overboard from ships at sea, have been picked up on the same shores, after safely traversing hundreds of miles of sea.

6. But by far the most constant and regular movement is one which, if we have never witnessed it before, cannot but fill our minds with wonder. Twice every day we find the edge of the sea to rise and fall, or to advance and retreat, upon the land. On a line of steep coast-cliff the surface of the sea may be observed slowly to sink during about six hours, and then for a corresponding interval slowly to rise again. On a flat or gently shelving shore this vertical movement manifests itself in another way. The fall of the water lays bare wide stretches of rocks, sand, or mud. During the retreat of the sea, we notice that successive ripples or waves advance less and less upon the beach. Step by step they seem to be drawn back, until in some large bays many square miles of flat ground are laid bare. During the advance of the water, on the other hand, wave after wave comes a little farther up the shore until all the uncovered flats are once more under the sea, and boats can sail over places which two or three hours before could be crossed on foot. This rise and fall of the sea is known as the Tides.

7. A little observation enables us to make quite sure that these daily movements have no connection with those of the air. When a strong gale is blowing, and driving the surface of the sea into foaming waves, it does not arrest the advance or retreat of the tide. The waves

continue to creep up or down the shore even in the face of the wind.

8. Again, we cannot but remark the wonderful regularity of the movements. Each advance of the water occupies rather more than six hours, and each retreat requires the same interval, so that the time of high or low water may be foretold for years in advance. From an early period of human history, and long before the real cause of the tides was understood, it was observed that there is a relation between them and the state of the moon. In the course of the observations which we have supposed ourselves to make, we find that the time of high water corresponds with the passing of the moon across the meridian, and that the highest and lowest tides occur at the time of new and full moon.

9. Putting now the sum of these observations together, we have ascertained the existence of three distinct kinds of movement in the water of the sea. First, wind-waves; second, surface-drifts or currents; and third, tides. From what was said in Lesson XV. it is clear that a fourth form of motion must be added to these, viz. a general creep or movement of the cold polar water along the bed of the sea towards the equator, and a flow of the upper waters towards the poles. We must consider each of these movements a little further.

10. (i.) **Wind-waves.**—These, like the ripples we send along the surface of a trough of water by blowing briskly at one end, are due to the pressure of the wind. Viewed from the top of a high cliff, or from the mast of a ship at sea, they may be seen to succeed each other in long parallel lines, travelling regularly and rapidly across the surface of the sea. Though the waves move forward, the water itself in the open deep sea shares but slowly in this movement. It rises and falls with a slight oscillation as each wave passes, though, if the wind continues to blow for a time, the surface water is gradually pushed onward. A field of corn or tall grass may often be

observed to be thrown into waves as a smart breeze sweeps over it, yet each stalk and blade retains its place, and merely bends up and down and to and fro with a kind of motion like that of the particles of water in a wave.

11. When the sea has been thrown into violent agitation by a storm, the waves do not at once subside when the storm is over, nor do they remain confined to that region only through which the storm has passed. So sensitive is the great body of oceanic water that the waves are propagated far beyond the area of the storm in vast undulations, or *ground-swell*, as this heaving is called. In the deep water of an open ocean the only trace of this movement is seen in the broad undulations which, like a great pulse, keep the surface regularly rising and falling. A ship sailing over this surface alternately mounts on the top of a swell and sinks into the trough between two of them. But where the sea shallows towards land, the up-and-down movement passes into a true onward rush of the water. The upper part of the undulation travelling faster than the lower parts, which are impeded by the friction of the bottom, begins to assume the aspect of a wave, curling over as it advances, like a huge wall of green water, to burst into foam against the land.

12. The motions of waves and ground-swell must be insensible at the bottom, in the deeper parts of the sea. In reference to the great body of the sea, they are merely surface agitations, probably seldom extending sensibly more than a few hundred feet downwards. So that when a hurricane is raising the surface of an ocean into the most violent commotion, we must think of the deep abysses below as dark, silent, and calm. And this is a matter of some importance when we consider the various offices of the sea, as will be done in the following Lesson.

13. Since the longer the space of deep sea over which the wind has to blow, the larger are the waves, we may

expect to meet with the grandest waves in the great oceans. In the accounts of storms at sea it is common to meet with such expressions as that "the waves were mountains high." The size of waves is apt, however, to be much exaggerated. Thus, a series of measurements, taken during a voyage across the North Atlantic, gave forty-three feet as the extreme height of the waves in stormy weather. In the narrow British seas, which are in a great measure barred off from the Atlantic, the largest waves are less than half the maximum size which they attain in that ocean.

14. No one can watch one of these huge waves as it nears the land without being impressed by the great force with which it beats upon the coast. The scene represented in the frontispiece of this volume may give some idea of the grandeur of a stormy sea. Each long undulation which rolls in from the outer sea mounts higher as it approaches the shore. At every step it becomes a more and more marked ridge of water, which begins to curl into a green-crested wave with a steep concave front towards the land. By degrees the upper advancing crest topples over, and the immense mass of water plunges with all its weight upon the rocks or sand of the coast-line. When the broken water has spent itself in rushing up the beach, it runs rapidly down again under the next advancing wave. In its recoil it drags back the sand and gravel with a loud rattle or hoarse roar, sometimes audible for many miles, as the stones of the shingle are ground over each other.

15. The force with which such *breakers*, as they are termed, fall upon the land has been measured and calculated. An undulation, or "roller," of the ground-swell twenty feet high has been computed to fall with a pressure of about a ton on every square foot of surface exposed to its reach. In summer the average force of the Atlantic breakers on the west coast of Britain amounts to about 611 lbs. on the square foot; in winter the force

is more than three times as great. On some occasions a pressure of as much as three tons and a half has been recorded. Some of the effects of the action of waves upon the land will be referred to in the next Lesson.

16. (ii.) *Currents.*—Since the sea responds so sensitively to every movement of the air, we may expect to find that as there is a regular system of circulation in the atmosphere, so there must be a corresponding system in the ocean. That this is really true has now been made clear by observations in all parts of the world. The sea has been found to be in continual circulation. Such evidence as that given in Art. 5 makes the reality of the surface-movement readily apparent. Those superficial parts of the sea which have a regular, continuous, onward motion, are termed "currents." They are called superficial because seldom more than 500 feet deep, which is but a small fraction of the depth of the total liquid mass over which they move.

17. The current-system of the sea presents in its leading features great simplicity. It arises from the action of the great atmospheric currents, which, steadily blowing on the surface of the sea, drive its waters before them. So that if we clearly follow the course of the circulation of the air, we need have little difficulty in understanding that of the sea. The general features of the circulation of the oceans are shown in Plate VIII.

18. The trade-winds set the surface-water of the sea in motion, so that it acquires a gradual tendency towards the equator. On the north side it comes from the north-east, on the south side from the south-east. Uniting along the equatorial belt, it necessarily assumes a westerly direction, and crosses both the Atlantic and Pacific Oceans as the *Equatorial Current.* In the former ocean it appears not to exceed about 300 feet in depth, and to move with a surface rate of not more than eighteen miles a day.

19. Were there no land to interfere with its flow,

this Equatorial Current would pass round the globe as a continuous stream of warm water. But owing to the position of the continents across its path, it is broken up and made to diverge in different directions, each of the independent branches receiving a distinct name. Thus in the Atlantic Ocean, setting out from the African coast, it crosses to the American side, where, striking against the projection of Cape St. Roque, it splits into two. The smaller branch, turning southward to form the *Brazil Current*, skirts the coast as far as the mouth of the La Plata, whence bending eastward, it once more crosses the Atlantic towards the Cape of Good Hope, and turns northward along the west coast of Africa, until drawn again into the great equatorial current. The larger branch sweeps round the northern coast-line of South America into the Caribbean Sea and the Gulf of Mexico, whence it issues through the Florida Strait as the warm and rapid ocean-river well known by the name of the *Gulf Stream*. After emerging from the narrows, with a maximum surface temperature of 80° Fahr. and a rate of 70 to 120 miles a day, this current runs parallel with the coast of the United States, but separated from it by a cold current, which descends from the Arctic seas. It gradually lessens in speed, and begins to spread out and to turn north-eastward. One portion, getting thinner and cooler, but yet retaining a higher temperature than that which is proper to the latitudes, and joining the general surface drift of the Atlantic water towards the north-east, extends to the coasts of Britain and Norway, and is said even to be traceable beyond Spitzbergen; the other turns southward between the Azores and the coast of Spain, bends round the north-west shores of Africa, and then coming once more within the influence of the trade-winds, is again sent across the Atlantic.

20. Within the wide circuit embraced by the second or south-eastern branch of the Gulf Stream lies a vast

area of comparatively still water, over the surface of much of which dense masses of sea-weed grow. This tract is known by the name of the *Sargasso Sea*.

21. In the Pacific Ocean the equatorial current has more room to develop itself. Starting from the western coast of the American continent, it streams westward across the whole breadth of that vast ocean, encountering on its way only the obstruction of scattered islands and submarine banks. It reaches the eastern margin of Asia where that continent is flanked by the islands of the Malay Archipelago. Hemmed in there, it divides into two main branches, one of which turns northward as the warm, rapid, and river-like *Japan Current*, and sweeps along the east coast of Asia as the Gulf Stream does along that of North America. The other and larger branch forces its way into the Indian Ocean, and joins the westerly equatorial current there.

22. These movements of the equatorial waters could not be carried on without drawing the water of the northern and southern hemispheres into the general circulation. Thus in the Pacific Ocean, which is open to the south, there is a general northerly set of the cold southern water. A broad current or drift pours into the hollow of the western coast of South America and, passing northwards, joins the equatorial movement. Part of it, however, strikes the promontory of Cape Horn, round which it sweeps as a well-marked current. Since the Pacific basin is almost closed at its northern margin, no great current of warm water can pass northwards through Behring's Strait, and on the other hand no considerable body of cold water can escape from the Arctic Sea, although a cold current does issue from the Strait and pass down the west side of North America. In the Atlantic basin, while the prolongation of the Gulf Stream bears its warmth far within the Arctic circle, the cold water of the Northern Ocean makes its way southward in return currents down each side of Greenland. The

L

more important of these currents descends from Davis Strait and pursues its course southward along the American coast, interposing between the land and the warm Gulf Stream, under which it partially passes. Icebergs have been carried across the Gulf Stream as far as lat. 36° S., which would seem to prove that the cold Arctic water continues to flow steadily with a sensible motion for a long way southward. The soundings of the *Challenger* show that the Gulf Stream between New York and Bermuda, with a temperature of 70° Fahr. and upwards, does not exceed about 70 fathoms in depth, and that below a depth of 600 fathoms the temperature of the water is always lower than 40° Fahr. Between the Gulf Stream and New York this cold water rises to within 300 fathoms from the surface, while farther north, along the shores of Nova Scotia, it comes quite to the surface, forming there the surface Arctic current out of Davis Strait.

23. But besides these surface-currents, which are always more or less readily perceptible, the waters of the ocean basins are affected likewise by a slower motion, which extends to the deepest abysses and from the equator to the poles. The existence of this general "creep," or slow diffusion, has been made clear by the temperature soundings described in Lesson XV. The very remarkable fact has been proved that even under the equator the main mass of the sea is cold, except a comparatively shallow layer at the surface warmed by the sun, and that at the bottom the temperature is as low as in the Antarctic and Arctic Seas. Were there no transference of cold polar water towards the equator, the temperature there ought to be comparatively high. But the presence of cold water even within 300 fathoms of the surface shows that there must be a deep and general movement from the polar regions to the equator, under the surface-currents, which, especially in the North Atlantic, carry the warm equatorial water into the cold

polar tracts. The cause of this movement of the cold water is not yet well understood.

24. (iii.) **Tides.** One of the most remarkable and regular of the motions of the sea was alluded to in Art. 6, as that alternate rise and fall of the water called the tides, the coincidence of which with the position of the moon could not but be noticed from the earliest times. To understand this motion, we must bear in mind that while all the members of the solar system mutually attract each other, there are two which specially influence our earth—the sun by reason of its vast size, and the moon on account of its nearness. Each of these two luminaries exercises a strong attractive force upon our planet, the tendency of which is to pull out the side of the earth which is opposite to it. The solid part of the globe does not sensibly yield to the strain, but the liquid ocean, unable to withstand it, is drawn outwards so as to be heaped up on that side where the attraction is exerted.

25. On the other side, too, where the distance from the attracting body is greatest, and the force of attraction therefore least, the water is not attracted so much as the earth, which is, as it were, pulled away from the water. Hence on that side the ocean likewise rises into another vast but less prominent swelling, the summit of which is exactly opposite to that on the near side of the earth. Were there no land to interfere with these motions, the distortion produced by this cause would give the form of an ellipsoid to the liquid mass covering our planet, its longer axis pointing to the centres of the earth and the moon. Were the earth and the other members of the solar system at rest, the two swellings of the ocean surface would remain stationary. But owing to rotation, these protuberant masses of water are forced to travel rapidly round the globe.

26. Let us fix our attention on that side of the earth which happens at the time to be facing the moon. Owing

to the earth's rotation, combined with the moon's own motion, our satellite appears to be revolving round the earth in about twenty-five hours. The outward bulging of the ocean-surface, which is caused by the moon's attraction, must therefore follow the moon, and run completely round the earth in about twenty-five hours, and the swelling on the opposite side of the earth must take the same course. Each of these two uprisings is felt, at every place which it passes, as a broad undulation or wave which appears once a day, or more exactly in every 24 hours 54 minutes, that being the length of the interval between each appearance of the moon on the meridian. Thus in that interval there are two times of *high-water* or *flood*, and two times of *low-water* or *ebb*.

27. It is not, however, the moon alone by which the water of the sea is affected in this way. The sun, too, draws the ocean-surface outward; but this influence, which has about one-third of the force of the moon's, is combined with the latter in the production of the broad tidal undulation. When the sun and moon are in the same line, which happens at new and full moon, their combined attraction must evidently most powerfully draw out the surface of the sea, while at the quarters of the moon their respective influences partially neutralise each other.

The accompanying figure (Fig. 19) shows how this effect is accomplished. The highest rise and, of course, the lowest fall, of the tidal undulation must occur at new and full moon. These are called *spring tides*. The least rise and fall take place at the time of the moon's first and last quarters. These are known as *neap-tides*.

28. Since the movements of the tidal undulation depend upon those of the earth, sun, and moon, they can be measured and predicted a long while beforehand. Observations have likewise been made in all parts of the world regarding the times of high-water and the height of the tide, so that the length of time taken by

the tide-wave to travel from point to point has been carefully determined. The rate of motion depends upon the depth of water and the absence or presence of land, being most rapid where the water is deepest and has no obstructions in its course. In the equatorial parts of the Atlantic Ocean it exceeds 500 geographical miles

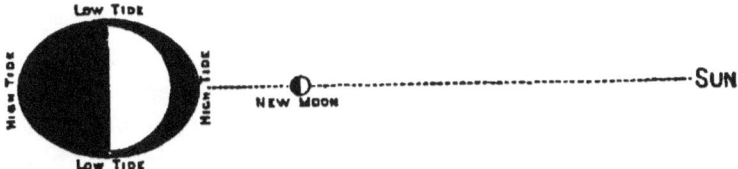

SPRING-TIDES

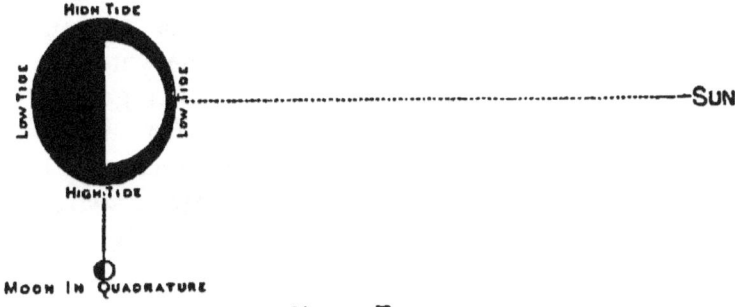

NEAP-TIDES

FIG. 19.—Diagram illustrating the origin of the tides.

in the hour. It is estimated to take fourteen or fifteen hours to come from the south of Africa to the southwest of Europe, where it gets into comparatively shallow water and requires to struggle through many narrow passages. Striking the south-west of Ireland, it branches out, one part turning north along the west coast of Ireland and Scotland, then bending south along the east coast, while the other part turns to the east up the English Channel. Now the former of these branches takes nineteen hours to circle round the British Islands as far as the mouth of the Thames, where it meets and mingles with the wave which started twelve hours later, and worked its way in about seven hours through the English Channel.

29. Though the tidal undulation is called a wave, and travels with great speed across the deep ocean-basins, the water itself does not partake of this onward motion. Each particle of water merely oscillates as the rapid pulsation passes, like the stalks of corn in a field, (Art. 10). But when the undulation enters a narrow and shallow sea, it becomes jammed between the converging coasts, and experiences increasing friction against the bottom. Its rate of motion consequently slackens, but in proportion as this takes place, the undulation gathers height and force, and becomes a true current.

30. In the deep ocean-basins, the passing of the broad tidal undulation probably produces little or no sensible effect upon the bottom. The surface of the water merely rises gently for about six hours, and then gently falls again, the total height of the undulation, from low-water to high-water, not being more than a few inches. Thus, in the middle of the Pacific Ocean, the rise is sometimes less than a foot; in the Atlantic, round St. Helena, about three feet. But where the undulation meets with the resistance of converging masses of land and a shallowing bottom, it is heaped up, sometimes, as in the Bay of Fundy, to a height of seventy feet, and rushes along as a great wave or as a surging and foaming ocean-river. In the Bristol Channel, which opens towards the west, the tide rolls up a rapidly-narrowing channel, and during spring-tides attains a height of forty feet in the estuary of the Severn. The advance of the tide is shown by a wave nine feet high, which rushes in front, succeeded by smaller ones, until the full tide has filled the estuary. This wave, or **bore** as it is called, forms a marked feature of the flowing tide in many bays and estuaries which open out towards the quarter whence

the tidal undulation comes. Thus, on the west coast of Europe it occurs in the Elbe, Weser, Seine, Dordogne, and Garonne. At the mouth of the Seine the tidal wave enters with a speed of fifteen to twenty feet per second and a height of six and a half to ten feet. The bore on the Hooghly River rushes up with such force as to do great damage to shipping unprepared for its approach.

31. When the converging shores are not closed at the end, as in the case of an estuary, but open out again into a wider sea, the tidal undulation takes the form of a rapid, river-like current, or race. Heaped up on the one side, it attains a higher level than on the other side of the narrow passage, through which, therefore, it rushes with great force and speed. The British Islands, lying, as they do, on the margin of the Atlantic Ocean, present many examples, of which the strait called the Pentland Firth, between the Orkney Islands and the north of Scotland, may be taken as one of the best. Standing at high or low water on a headland overlooking that narrow passage, one looks down upon a strip of blue sea about six or eight miles broad, opening westward into the Atlantic and eastward into the North Sea. As soon as the tide begins to flow or to ebb, this smooth belt of water becomes more and more troubled, until, when the motion of the tide is at its height, it sweeps past at a rate of eleven miles in the hour, boiling up and foaming along like some vast river. Here and there, where sunken rocks lie in the path of the current, the tumult of the water is strongest, while, should a high wind be blowing against the tide, huge waves and sheets of foam are tossed high into the air, and the whole surface of the firth becomes white with breakers. No small vessel can then attempt to force its way through the strait.

32. In other narrow passages between islands where the tide is thrown from side to side against sunken rocks,

or where two opposing currents meet each other, the water forms **whirlpools**. Such are found in the well-known Corryvreckan between the islands of Jura and Scarba on the west coast of Scotland, where the current runs alternately east and west at a speed of eleven miles in the hour. The famous Maelstrom, at the southern end of the Lofodden Islands on the Norwegian coast, is another illustration.

33. The strip of sand, gravel, or mud which is alternately covered and laid bare by the rise and fall of the tidal undulation is called the **beach**. When the tide is at the full, the sea reaches to the upper limit of the beach or **high-water mark**, when the tide is at the ebb the

Fig. 20.—Diagram showing the relation of the beach to the lines of high and low water.

sea does not come farther than the lower limit of the beach or **low-water mark**. The distance between these two limits must evidently depend partly upon the height of the tides, and partly upon the slope of the beach. It is, of course, greatest where the largest rise and fall of the tides is combined with the gentlest inclination of the shore.

LESSON XVIII.—*The Offices of the Sea.*

1. In the foregoing Lessons the more important purposes which the sea serves in the general life of the earth have been described. Let us here consider them in a brief summary.

2. (i.) The sea supplies most of the moisture of the atmosphere. Were the air overlying the land dependent for its moisture solely upon the evaporation from the land, rain and dew would almost cease, and neither vegetable nor animal life could flourish. The clouds that overspread the sky and discharge their rain-showers upon the ground, the springs that feed the brooks, the innumerable streams that go to swell the bulk of the great rivers, the snows that overspread the higher mountains, the dews that refresh the surface of the earth even in tracts where no rain falls, all come directly or indirectly from the sea. In spite of the enormous volume of fresh water continually poured into the sea by innumerable streams, no change of the sea-level is perceptible. The water is evaporated again, rises into the air and is borne along by the winds, till it is condensed and once more discharged upon the land.

3. The influence of the sea upon the moisture of the air is well shown by the much greater rainfall of maritime than of inland tracts (Lesson X. Art. 29). The air lying over the sea is probably for the most part not far from the point of saturation, so that when it moves away landwards it is ready to part with some of its vapour as soon as it meets with a mass of air or of land colder than itself. Near the coast of India the evaporation from the sea is said to amount to about three-quarters of an inch in the twenty-four hours, or nearly twenty-three feet in the year. A rough estimate gives a depth of fifteen feet as annually evaporated from the trade-wind region of the oceans.

4. In the torrid zone, where the evaporation is greatest, much of the moisture raised from the sea undoubtedly falls back into it in those heavy and continual rains which characterise that belt of the earth's surface. Some of it goes to supply the copious rainfall of such regions as the Khasi Hills (Lesson X. Art. 31), the Himalaya,

and the high lands of Abyssinia. Very little of this equatorial vapour, however, can pass over to the temperate regions on either side, because the air which contains it, after discharging abundant showers in the zone of constant precipitation, ascends into the high regions of the atmosphere as a cold and comparatively dry current, travelling away from the equator, and reaching the surface of the earth again, ready rather to take up moisture than to part with it.

5. In Europe the rains are supplied chiefly by the moist winds that drink up the vapour of the warm surface of the Atlantic; west, and especially south-west winds are wet, east and north-east winds are dry. The vast cauldron of the Indian Ocean furnishes the rains of the south of Asia and the east of Africa. When the south-west monsoon blows, it bears the vapour of that ocean to India and China, and discharges it in deluges of rain. The large rivers of western Africa are supplied by the monsoons, which bear the vapour of the Atlantic into the interior of that continent. In North America the mountainous western sea-board derives its moisture from the moist winds that blow from the south-west across the Pacific Ocean. In South America the south-east trade-winds carry the vapour of the South Atlantic across the continent, until they discharge the last of it on the slopes of the Andes, and pass over to the Pacific coast as dry, rainless winds.

6. (ii.) **The sea regulates the distribution of temperature.** Its currents carry the heat of the tropics to warm the regions lying towards the pole; while, on the other hand, they bring the cold of the poles to temper the heat of lower latitudes.

7. This general influence of the sea, well illustrated by the charts of isothermal lines, has been referred to in Lesson IX. Art. 15, and the cold and warm currents of the North Atlantic ocean were given as examples. The heated water of the Gulf Stream turns north-east-

ward and crosses the Atlantic, spreading out over a larger surface and losing heat as it advances, yet distinctly traceable by its warmth even into the Arctic seas. The average winter temperature of the sea round the British Islands is considerably higher than that of the land. But as already remarked (Lesson XV. Art. 2) it is not directly by contact with the land that the sea tempers the climate, but by convection currents of air. It warms the air overlying it, which then passes on and carries the warmth across the land. And here we see the advantage to the west of Europe in that spreading out of the Gulf Stream water which has been referred to. Were this current to pass across the ocean with the same breadth and speed which it has when it issues from the Florida Strait, even though it should retain its highest temperature, it would be too narrow to produce much effect on the climate of Western Europe. But by widening out over a far larger area and moving much more slowly, it exposes a much greater surface to the south-westerly breezes, which, warmed and moistened by it, pass over to Europe.

8. The extent of this amelioration of the climate in Western Europe may best be seen by comparing the winter temperature of places on the same latitude on opposite sides of the Atlantic. Hammerfest, the most northerly seaport in the world, is open even in January, while to the west, on the same parallel of latitude, the east coast of Greenland is covered with snow and ice, and the sea remains frozen in vast floes even during summer. Still farther south, the harbours of Glasgow and Liverpool are not only never frozen, but enjoy a mean annual temperature of 47°—51° Fahr. On the corresponding latitudes in North America the coast-line of Labrador is cumbered with ice all the year. Even at St. John's, Newfoundland, which is two degrees farther south than Liverpool, the harbour has been closed with ice in the month of June.

9. These contrasts, which are probably the most striking of the kind anywhere to be found upon the globe, do not depend wholly upon the influence of the warm Atlantic current in raising the temperature of Western Europe. They are partly caused also by the influence of the cold Arctic current, already described as descending from Davis Strait, keeping close to the American coast and depressing the temperature there. Even as far south as lat. 44°, that is, on the same parallel as the south of France and north of Italy, the water that washes the shores of Nova Scotia is not more than five or six degrees above the temperature of thawing ice. A further cause of the low temperature of Labrador and adjacent parts of North America is to be sought in the low atmospheric pressure over the North Atlantic, and the consequent flow of cold air from the north-west.

10. In the southern hemisphere, where sea so largely predominates, the climates are distributed with comparative regularity according to latitude. In the northern hemisphere, however, where the equalising influence of the sea is opposed by the existence of large masses of land, which are alternately heated and chilled, no such regularity exists. And yet even there, the tempering action of the sea makes itself sensible along the borders of the land, where neither are the winters so intensely cold nor the summers so hot as they are inland. Hence a **continental climate** is one of extremes—great heat in summer, severe cold in winter; an **insular climate** is more equable—a distinction well brought out by comparing the summer and winter temperatures of Ireland with those of regions on the same latitudes in the midst of the continent, such as the heart of Russia. (See Lesson XXXI.)

11. (iii.) **The sea wears away its shores, and thus tends to reduce the area of the land.** No one can watch the action of the waves (Lesson XVII. Art. 14) without being well able to realise how powerful

it must be in the destruction of even the most rocky coast-line. The mere weight of so many tons of water, as are suddenly thrown by a huge wave against the land during a storm, suffices to loosen and detach fragments of rock. Even the hardest rocks are apt to be split up by the action of the atmosphere (Lesson XXIX.), and consequently the waves are aided in their work of demolition by the cracks and lines of weakness which the atmosphere has prepared for them.

12. Fragments of stone detached from a sea-cliff become powerful implements in the further destruction of the coast-line. They are caught up and hurled forward by the waves, battering down the cliffs, and at the same time being further broken up themselves. Ground to and fro by the advance and recoil of the waves, they assume the smoothed and rounded forms so familiar in the gravel and shingle of the sea-shore. But even then, their demolition continues, for they are still rolled backwards and forwards, until they become at last mere sand and mud, in which condition they may be swept away out to sea, and laid down in the quiet depths there.

13. There is yet another way in which the sea loosens and demolishes the rocks on its margin. Those who live at a seaport, on any coast exposed to storms, must often have noticed that after a violent gale portions of the solid masonry of the pier-walls and breakwaters have been started from their places. Even with the utmost care in the building and repair of these works, it is hardly possible to avoid injury of this kind; and unless constant vigilance is used in cementing every crack and opening in the masonry, the destruction of the pier or bulwark may be rapid and complete. Now it is not by the mere weight of water thrown against the building, nor by the impact of any blocks of stone hurled forward by the waves, that the stones are started out of a piece of well-built masonry. When a large wave falls upon such a surface, the air in every cranny of the wall is driven

inward. When the wave recoils again, a great suction takes place behind it, and the compressed air rushes out of the masonry. Moreover, where the water is driven in with immense force so as to fill crevices and openings within the masonry, it exerts upon their walls the same pressure as the wave of which it is for the moment a part, on the principle of the hydraulic press; and this pressure may amount to three tons on every square foot. After a time some weak part of the work is discovered; one or more stones are loosened and then pulled out at the recoil of some large wave, so that a breach is made, which, if the storm continues, may be rapidly widened. The same kind of influence is exerted upon cliffs. Every mass of solid rock is more or less traversed by crevices, which afford scope for the combined action of compressed air and waves. When the sea has drilled a cave at the base of a cliff, the action of the air, alternately compressed and released as the waves rush in and out, loosens the rock at the end and roof of the passage, so that if the cliff be not too high, or the rock too solid, an opening is actually made between the end of the cave and the ground behind the top of the cliff. Through this opening, called a *blow-hole*, the sea in storms sends clouds of spray. Sometimes the mouth of the opening lies a short distance from the edge of the sea. It is singular to find on a hill-side, or in the middle of a field or moor, a dark cauldron-like hole, with the sea ebbing and flowing at the bottom.

14. Since even cliffs of the most solid rocks are worn down by the incessant attacks of the waves, we may readily believe that where the material is of less durability the rate of destruction must be still more rapid. On some parts of the eastern shores of England, where the waves beat against cliffs of crumbling clay, the rate of waste occasionally reaches to as much as the loss of a yard (here and there even five yards) of land in the year. The sites of some of the former ports and towns

of Yorkshire now lie beneath the restless waters of the North Sea, a mile or more from the present shore.

15. We must not suppose, however, that the sea is everywhere cutting away the margin of the land. It does so chiefly on those coasts which lie exposed to prevalent storms, and where the form of the shore allows

Fig. 21.—View of the sea-cliffs south of the river Tyne (magnesian limestone worn away by the waves, and the isolated fragments left). See also Fig. 76.

the waves to act with effect. But on every coast-line there are sheltered places where the waves either do not encroach at all, or at least do so very slowly. For example, if the sea is shallow for a long distance outward, with a flat, sandy beach and a low coast-line, the largest waves may roll in and spend their force merely in rushing along the flat beach, and may do no damage to the shore. But in many places the sea, instead of removing material, casts it up on the land. The finer sediment swept away from one part of the coast and carried off by the currents is sometimes thrown ashore again at no great distance. When that takes place, the land may gain there as much, or nearly as much, as it loses from that part of its margin whence the sand and mud are derived. In this way the flat shores of Lincolnshire encroach upon the sea, because they continually receive some of the material removed from the coast of Yorkshire. In many parts of the world, for example along the coast of North America from the borders of Mexico to those of Virginia, and on both sides of the Madras Presidency, so much mud and sand are brought down by the rivers from the land that bars are formed, by which the land behind them is protected from the waves.

16. (iv.) **The sea receives and preserves the materials out of which new land will in course of time be formed.**—We shall trace in a later Lesson how the surface of the land undergoes continual decay. Even its hardest rocks crumble down, and their remains are swept out by brooks and rivers into the quiet depths of the ocean, where, at no great distance from the shores, the sand and mud derived from the decay of the surface of the land are laid down. There they remain undisturbed, slowly accumulating and spreading, until some future movement of the earth's crust shall raise them above the level of the sea.

17. Many of the rocks of which the present dry land is formed contain abundant remains of coral, shells, fishes, and other marine creatures. These animals lived in the sea, and when they died their harder parts, enclosed in the sediments of the sea-bottom, were preserved there; thereafter these sediments, slowly consolidated into rock, have been upraised so as to form the land on which we live. This has happened frequently in the past, and will probably continue in the future.

18. Besides the deposits formed of the fine sand and mud carried away from the land, vast spaces of the bottom of the ocean are likewise covered with a slowly-gathering deposit, derived from the remains of minute

forms of life (Lesson XIV.). Some of the rocks of the land, the chalk of England and France for example, have been laid down in much the same way, by the gradual accumulation of the tiny shells of foraminifera and other inhabitants of the sea.

19. It is evident that while the surface of the land is subject to continual decay and removal, that portion of the solid earth lying below the comparatively shallow depth (a few hundred feet at most) to which the influence of waves extends, is preserved. By far the largest part of the destruction which the sea works is done between tide marks. (See Fig. 21.) Waves, as we have seen, do not stir the deeper parts of the ocean, which are therefore left in undisturbed repose. When we reflect how large a proportion of the earth's surface is covered with sea, we perceive that although the waves are always beating against the shores, the destruction which they effect takes place only on the mere margin of the land, while the vast spaces under the sea are not only preserved from waste, but are continually receiving fresh deposits. So that if we consider the action of the sea as a whole, we find it to be far more preservative than destructive.

CHAPTER IV

THE LAND.

Lesson XIX.—*Continents and Islands.*

1. From the two envelopes of air and sea, in which our earth is enclosed, we now pass to the consideration of what lies beneath or within them—the solid mass of the planet. One of the first points to be noticed here, is that while both air and sea admit of being penetrated and examined all over the world, only a very small proportion of the solid framework of the globe rises out of the sea to form land, and of that comparatively insignificant part little more than the mere surface or skin is known to us. We cannot at will pierce the land as we can sound the sea. Evidently, therefore, all that can be discovered by actual observation regarding the constitution of the solid globe must be derived from what may be seen at or near the surface of the land.

2. What then, to judge from all that can be learnt from its visible parts, is this solid globe on which we live? We are familiar with the earth and soil of the surface, and we have seen the rocks beneath it—in one country clays and limestones, in another sand and gravel, in another sandstones and shales, in still another granites and other crystalline formations. How have these various materials arisen? Can we learn anything regarding their origin, and do they in any way throw light upon the history of the earth? Why, for example, should the land rise into lofty ridges at one place and sink into low plains

in another; and have these ridges and plains been always the same as now?

3. No matter in what part of the world we may live, such questions as these must often arise in our minds. The more prominent the features of the land, the more, of course, do these questions press for an answer. In a mountainous country, for example, or one from which a distant range of mountains can be seen, the high peaks and ridges of the land form so striking a contrast to the level plains below, that we cannot help asking how and when these vast elevations took their rise. We watch the clouds on the far summits as they gather into thunderstorms, or dissolve into showers of rain among the valleys, or powder with fresh snow the white shoulders and crests of the mountains. There seems to be a constant turmoil in the air of these heights. What effect must it have upon the forms of the mountains? Do these ever recurring tempests, these frequent rains and snows, which fill the torrents and feed the great rivers, leave the outlines of the mountains unchanged?

4. But even where the landscape is least marked by any striking feature, an observant eye cannot fail to notice much that is suggestive of inquiry. No scene, for instance, can be more monotonous than that presented by the wide plains at the mouths of large rivers. And yet, when the mind is once opened to perceive the interest of such subjects as these, can the nature of the soil of these flat lands escape notice, and fail to be connected with the mud brought down by the river? Whence does the mud come? How long has the river been busy in strewing it over these plains?

5. Such are some of the questions that are now to come before us in the Lessons contained in this chapter. In entering upon them it is well to remember that, as far as possible, we should try to find illustrations of them in the district wherein we may chance to live. We may not be able to do so in all, nor even perhaps in most cases.

Yet the search for examples will always be of the utmost use in making us better acquainted with the physical geography of our country, and in accustoming us to the observant use of our eyes. The features of a mountain-chain, a plain, a table-land, a watershed, a spring, a brook, a river, a lake, and of every other part of the land, should be studied on the ground, wherever we can find examples of them. In all such cases, too, we should not rest content until we have put to the proof the statements regarding these features in the Lessons, and seen how far they are borne out, or can be extended by what we see in nature for ourselves.

6. As the land consists of those portions of the solid globe which happen to project above the level of the sea, its distribution shows the position of the large wrinkles into which the surface of the planet has been folded, while its form and its materials must evidently furnish our chief sources of information regarding the composition and the history of the general mass of the earth. First, the **distribution** of the land over the globe may be noted; then its horizontal outlines or **coast-lines** where it meets the sea; next, its vertical outlines or **relief**, as it rises up into the air.

7. **Distribution of the Land.**—The total area of dry land on the surface of the globe has been computed to amount to 52,000,000 of square miles. By much the larger part of this land lies in the northern hemisphere. (Lesson V.) Within the Arctic Circle (Plate I. Fig. 5) it forms an almost continuous ring round the north polar regions, whence it extends southwards in long irregular masses which taper away into points (Lesson V. Art. 10). Each of these masses of land or **continents** may be looked upon as consisting of a northern and southern division. In the western hemisphere they are united by a narrow neck of land or **isthmus**. In the eastern hemisphere, Europe and Africa are separated by the narrow water-channel or inland sea

called the Mediterranean. Europe and Asia form, indeed, one united and continuous continent, which is prolonged southwards by Africa on the one side, and by a vast **archipelago**, or clusters and chains of islands, into Australia, on the other. In a general view, the land may be regarded as forming three pairs of continents: the most regular being North and South America, and the least regular being Asia and its insular prolongations into Australasia.

8. Looking now at the position of these continental masses we may observe that there is among them a general tendency to run in a north-west and south-east direction. This is specially noticeable in the case of America. It is seen, too, in the bend of the islands from the south-eastern margin of Asia to Tasmania and New Zealand. Even in the European and African pair of continents it may be traced, in the plateau or ridge that connects Greenland with Iceland, the Faroe and British Islands and the main mass of Europe. An uneven and broken ridge of land is thus seen to extend from the polar circle south-eastwards to the Cape of Good Hope. Hence the shape of the Atlantic ocean is determined as that of a deep long channel or trough between the Old and New Worlds. It will be noticed too that the central submerged ridge of this ocean runs in a general sense parallel to the land on either side. Owing to the expansion of Asia to the north-east, the American and Asiatic continents almost meet at Behring Strait, thus enclosing the vast Pacific basin on the north.

9. In order to have a clear notion of the main heights and hollows on the surface of the solid globe we cannot merely consider the trend of the continents. Besides these chief masses of land, innumerable smaller patches project above the surface of the oceans, in the form of scattered groups or clusters of islands. As these are only the tops of submerged ridges or mountains, their position serves to indicate the direction of the submarine

elevations of the earth's surface, and the submerged parts of the continental areas. Much information regarding the form of the sea-bottom and its relation to the land has been gained from deep-sea sounding, as already remarked (Lesson XII.). Such a map as that on Plate VII. is hardly less suggestive in regard to the continental than to the oceanic areas. Turning now more particularly to the land itself, we observe it to be defined by two limits, one where it meets the sea—that is, its coast-lines, the other where it is surmounted by the air, giving its vertical relief or contour.

10. Coast-lines of the Continents.—Though the limit between land and sea is always sharply drawn, the map of the world shows how varied and irregular it is. But even the largest and most detailed map can show only the more prominent curvings and windings of the coast-line. When such a map is carried along the edge of the land, and compared with the actual boundaries of the sea, many little projections and recessions of the coast are seen which, from their small size, could not find a place upon the map. Comparatively seldom does the coast continue in a straight line for more than a brief space. It curves to and fro, jutting out into *headlands*, *capes*, or *promontories*, and retiring into wide *gulfs* and *bays*, or into narrower *firths*, *creeks*, and *inlets*.

11. Besides these abundant changes in its horizontal extension or plan, the coast-line of the land presents continual variety in its vertical elevation or relief. Sometimes, for example, it rises abruptly out of the sea, in steep and lofty cliffs, against which the waves are ever beating. Elsewhere, it shelves gently into the water. The slope of the land above the surface of the sea is, for the most part, prolonged under sea-level. Hence, where the coast-line is precipitous, the water is usually deep (Fig. 22), while, on the other hand, a low shore commonly indicates shallow water (Fig. 23). This relation

between form of ground and depth of water is so general, that in navigation a vessel will usually be kept well away from a low shore, though brought without hesitation close under a high headland.

Fig. 22.—Steep shore descending abruptly into deep water.

12. As a rule the projecting and higher parts of a coast-line consist of harder materials, the receding and lower parts of softer materials. This is true on a small

Fig. 23.—Low shore shelving into shallow water.

scale of any exposed piece of shore, and likewise of the larger irregularities of the coast all over the world. Wherever a bold headland stands well out to sea, we may infer that it consists of some hard rock which can offer a stout resistance to the waves. Wherever, on the other hand, the land is shown on the map as retiring into wide unindented bays, it may be presumed to be low and to be formed of soft clays or other materials that have yielded more easily to the encroachments of the sea. In the great majority of cases these inferences, drawn merely from the examination of a map, would be correct.

13. Compare now the coast-lines of the three pairs of continents. A remarkable contrast is at once observable between those of the northern and southern divisions. The northern continents are marked by coast-lines so indented that the oceans penetrate far inland in many bays, creeks, inlets, firths, and inland seas. The southern continents, on the other hand, are distinguished by long monotonous stretches of shore, unbroken by bay or creek. The following calculation has been made of the relative proportions of coast-line to extent of surface among continents.

FIRST PAIR—
  N. America has 1 geograph. mile of coast-line to 265 sq. miles of surface.
  S. America ,, 1 ,, ,, ,, 434 ,, ,,

SECOND PAIR—
  Europe ,, 1 ,, ,, ,, 143 ,, ,,
  Africa ,, 1 ,, ,, ,, 895 ,, ,,

THIRD PAIR—
  Asia including the Islands ,, 1 ,, ,, ,, 469 ,, ,,
  Australia ,, 1 ,, ,, ,, 332 ,, ,,

14. It will be seen that the contrast is most marked between Europe and Africa. The former continent has six times more coast-line in proportion to its superficial extent than Africa has. There can be little doubt that this contrast has had a powerful influence on the civilisation of the two continents. In the one case abundant bays and arms of the sea have offered ready facilities for discovery, conquest, and commerce. Nation has been brought into contact with nation; the arts of peace and war have spread into every country; no region, however remote, has lain wholly beyond the reach of communication with the rest; and hence all the communities of Europe, whether slowly or rapidly, have been carried

forward in the general progress of mankind. But contrast this abundant intercourse and steady advance with the condition of Africa, its vast expanse of hitherto inaccessible and unopened territories, peopled by races having no intercourse with other tribes save their immediate and usually hostile neighbours, and remaining from time immemorial in the same stationary condition of barbarism. The noble rivers and lakes of Africa, however, will doubtless eventually become highways of civilisation into the very heart of the continent, and be made to serve the purpose which in old times was accomplished in Europe by the far-reaching arms of the sea.

15. **General Relief of the Continents.** — The horizontal outlines or coasts of the land are greatly surpassed in variety and interest by its vertical outlines or relief. It is easy to show the former upon a map, but much more difficult adequately to express the latter. Hence even the best maps convey a most imperfect idea of the external aspect of the land. Leaving details for the next Lesson we may notice at present some general features which are illustrated in all parts of the globe.

16. Even the loftiest peak of land rises but a small way above the average level of the earth's surface. Think, for instance, of the proportion between the height of even the highest summit of the Himalaya Mountains and the total bulk of the earth. The top of Mount Everest stands 29,000 feet, or $5\frac{1}{2}$ miles, above the level of the sea, an enormous height when compared with the little undulations of the plains or even with the heights of many hilly countries. And yet it is only $\frac{1}{1436}$th of the polar diameter of the earth; so that on a globe of ten feet in diameter the highest mountain in the world would be represented by a little projection only one-twelfth of an inch in length.

17. If, then, the most stupendous mountains are really so small, it is evident that the whole mass of dry

land must form but an insignificant part of the entire bulk of the earth. For by far the largest part of the land is not mountainous. Calculations have been made regarding the probable average height of the land, if the mountains could be levelled down and the valleys filled up so as to reduce the whole to one general level. This has been variously estimated to be from 1000 to 1450 feet. The continents, however, differ much from each other in this respect. Europe has been computed to have an average elevation of 973·6 feet. That of Africa must be more than twice as great. Vast mountain-chains do not form such a large proportion of the mass of the land as they might be supposed to do. It has been calculated, for instance, that if the whole of the Alps could be ground down and spread over Europe, the surface of that continent would not be raised more than about 21 feet.

18. In no respect does the land stand out in more marked contrast with the sea than in the endless variety of its relief. While the surface of the oceans is one vast level, that of the land exhibits every contrast of form, from the jagged peaks and precipices of the mountains down to the level flats of the meadows and plains. We are apt to see no order in this variety. Mountain and valley seem to succeed each other, and to change their shapes at random. And yet an attentive study of them shows that they have not been stuck down on the earth's surface at mere hazard, but have an intelligible meaning and history, and a traceable connection with each other.

19. First of all it is to be noticed that each of the continents is traversed by a line or axis from which the ground slopes on either side to the sea. This axis does not necessarily coincide with the highest parts of the land: these may lie now to the one side of it, now to the other; but it is the line of average highest elevation, as is shown by the way in which the rivers flow from each

side of it. Nor is the axis placed down the centre of the continent; more usually it is much on one side. In America, for instance, it lies close to the Pacific seaboard. In Europe it runs from Cape Finisterre through the chain of the Pyrenees, the Alps, the Carpathians, and the Caucasus to the shores of the Caspian Sea.

20. The position of the axis determines the general slope on either side. When it runs along the centre of a continent the average angle of slope on either side will be the same. When it lies close to one side the angle must be higher on that side than on the other. Each continent or country, with an axis lying far from the true centre of the region, has therefore a short and steep slope on one side, and a long and gentle slope on the other. South America presents the most remarkable example of this feature. The axis, with an elevation of perhaps 8000 or 10,000 feet, runs down the line of the Andes at a distance of only from 50 to 100 miles from the Pacific, but 2000 miles from the Atlantic Ocean. On a much smaller scale the same character is shown by Scandinavia, where the axis lies close to the Atlantic, and where consequently the ground descends rapidly into that ocean from the snow-fields of Norway, but slopes gently eastward across Sweden into the Gulf of Bothnia. In the British Islands, too, the western slope is short and steep in the northern half of Scotland and in Wales, while the eastern slope is long and gentle.

21. But observe how the axial line runs through the great continents. On the whole it is so placed that the steep and short slope shall face the larger ocean. Look again at America, with its vast line of heights towering steeply out of the Pacific on the one hand, and inclining gently for hundreds of miles into the Atlantic on the other. On the opposite side of the Pacific the lofty highlands of Asia rise as a great mountainous girdle which stretches from Kamtschatka to Arabia, and is thence prolonged through Africa to the Cape. The slope to the

Pacific and Indian Oceans from that range of high ground is comparatively short, while on the opposite side to the Arctic and Atlantic Oceans it is gentle and long. It would seem that on the average the gentler slope of a continent is four or five times longer than the steep one.

22. Moreover, it will be noticed that subordinate to the main axis a minor axis or ridge is apt to occur on the other side of a continent fronting the smaller ocean. This likewise may be seen most conspicuously in America. The line of the Alleghany and White Mountains lies near the Atlantic border in North America, and the Sierras of Brazil rise close to the sea in South America. These dominant lines cannot be mere accidents; they must be due to the same causes which have ridged up the whole of the solid land.

23. Again, it is evident that when the continents are flanked with high ground on either side they must contain wide plains, or at least vast spaces of lower ground, in their centre. America illustrates this aspect by the great plains of Canada, of the Mississippi, Orinoco, Amazon, and La Plata. The greater part of Europe is a low plain bounded by the chain of the Alps, Carpathians, and Caucasus on the south, the Ural Mountains on the east, and the high grounds of Scandinavia and Britain on the west.

24. In most cases the upheaval of the continents has been accomplished in such a way that these central plains or lower grounds slope seawards. The water which falls upon them, therefore, is not retained there, but flows off to the ocean. In the crumpling and erosion of the surface of the earth, however, depressions have been formed upon each of the continents below the level of the surrounding country. Into these the water flows, filling them up till they overflow and discharge their surplus by means of streams. But in desert regions the rivers gradually disappear in the sandy wastes, or if they flow into a lake, there is no outflow. In such places the evaporation is at present so great as to balance the supply of water. On the high central platform of Asia a region of this kind, full of salt-lakes, lies enclosed between the Hindu Kush and Thian Shan Mountains, and stretches through Turkestan and Mongolia for a distance of about 2000 miles. In North America a much smaller basin, shut in among the ridges to the west of the Rocky Mountains, receives its drainage into lakes without outlets, of which the Great Salt Lake of Utah is the most important. In the south-east of Europe the depression of the Caspian descends far beneath the level of the sea; even the surface of that inland sheet of salt water being about eighty-four feet below the general level of the oceans. The surface of the Dead Sea, which fills a small detached hollow shut in by high grounds, is 1298 feet below the level of the Mediterranean.

We have now traced the position and broad general outlines of the great ridges on the surface of the earth. The following Lesson deals with some of the more prominent features of the surface of the land.

LESSON XX.—*The Relief of the Land—Mountains, Plains, and Valleys.*

1. Although the surface of the land is very uneven, rising sometimes, as in the Himalaya, five and a half miles above, and sometimes, as in the hollow of the Caspian (3000 feet deep), sinking nearly three-quarters of a mile below the sea-level, its inequalities are not altogether disposed at random. Each continent has its own system of heights and hollows, and these are grouped in relation with the general axis.

2. The main features of the continents are the lines of **mountain-chains.** It is along these lines that the elevation of the land has been greatest. They are, as

it were, the crests of the great waves into which the surface of the solid globe has been thrown. They govern the trend of the chief valleys, they determine the position of the plains, they regulate the climate, the winds, rains, and rivers of the land. They ought first to be considered, therefore, in any description of the general aspect of a country: we may then pass to the valleys, plains, and table-lands.

3. **Mountains.**—The term "mountain" is rather vaguely used to describe any large and lofty elevation rising conspicuously above the region which surrounds it. Sometimes a single conical mountain towers above a plain or shoots up from the sea. Solitary cones of this kind are usually volcanoes, such as Etna, Vesuvius, and the Peak of Teneriffe. More frequently a mountain is merely a more prominent part of a long lofty ridge, and is joined at the sides to other similar mountains, into which the ridge is divided by the valleys that cross it. Such a connected series of mountains is called a **mountain-range**. It often happens that two or more such lines of mountain run parallel with each other as one vast and continuous **mountain-chain,** or **mountain-system.** Let us consider for a little the aspect of some leading mountain-chain as an illustration of this feature of the surface of the land.

4. The Alps of Europe have been so long familiar to mankind that their name has passed into use as an appellation for the main mountain system of any country. To one who approaches it from the north, that noble chain of mountains only reveals its character step by step. Crossing the plains of France or Germany into the district of the Jura, the traveller finds the ground begin to rise and sink in long ridges and valleys, which follow each other in parallel lines, like undulations on the surface of a great ocean. The smooth valleys are green and fertile; the ridges, for the most part also smooth, bear their slopes of rich pasture and shaggy

wood, but often show along their crests long narrow valleys or amphitheatres bounded by walls of rock. Here and there deep transverse gorges cross the ridges and carry the drainage out into the plains beyond. Threading his way through these transverse valleys, the traveller finds the ridges on either hand to grow higher, and the hollows between to become steeper and deeper, until he reaches one of the last and loftiest of these outer elevations. From its top he can see on the one side, as in a map, the succession of wave-like folds of

Fig. 24.—Section of one side of a Continent.

hill and valley across which he has journeyed; on the other, beyond a broad plain that lies at his feet, his eye may take in the whole panorama of the Alps—a vast array of mountains crowding behind each other along the sky-line, their higher slopes and crests white with snow, and hardly at times to be distinguished from the white clouds that rest upon them.

5. On a nearer view this great line of heights is found to be flanked with minor ridges, rising into loftier and bolder masses than the Jura, as they approach the main mass of the mountains, but still retaining their covering of forest, save on the steeper or more rugged slopes, where the naked rock forms broken crags or steep bare precipices. Beyond these outer bulwarks rise the central ranges, with flat meadows and corn-fields or deep still lakes at their base, dark pine-forests and green hollows of pasture on their sides, but rearing their shoulders above the limit at which tree or grass can live, and raising their heads far up into the region of eternal snow.

6. It is not until we have gained some commanding point of view above the snow-line that we can form any

proper conception of the vastness of the scale on which these mountain ridges have been piled up. The point which we have reached after some hours of laborious climbing seemed the very summit of a mountain when we stood in the valley far below. And yet we now find a vast sweep of much higher ground all around. Peak rises behind peak, crest above crest, with infinite variety of outline, and with a wild grandeur which often suggests the tossing and foaming breakers of a stormy ocean. Over all the scene, if the air be calm, there broods a stillness which makes the majesty of the mountains yet more impressive. No hum of bee or twitter of bird is heard so high. No brook or waterfall exists up among these snowy heights. The usual sounds of the lower ground have disappeared. Now and then a muttering like distant thunder may be caught, as some loosened mass of snow or ice falls with a crash down the slopes; or as the wind brings up from below in fitful gusts the murmur of the streams that wander down the distant valley.

7. It is seldom that these high summits are free from cloud during at least some part of the day. They rise so far into the air, that often for weeks at a time they are wholly wrapped in mist. The traveller who climbs them finds sometimes that the higher points rise above the cloud-belt into the clear upper sunshine, and he may even now and then look down upon a thunder-storm raging along the lower crests of the mountains and sending torrents of rain upon the valleys, while he has a clear blue sky above him.

8. Another feature that strongly impresses the mind is the contrast between the climate and vegetation of the plains and those of the mountains. Among the Swiss valleys, for instance, corn-fields, gardens, orchards, and vineyards cover the lower grounds, while thick woods climb the slopes. As one ascends, the plants of the valleys gradually disappear, others of species peculiar to

the higher and colder air of the mountains take their place; the pine-trees, after a time, grow fewer, and at last cease, until after farther ascent, the line of permanent snow is reached, at a height of about 9000 feet above the sea (Fig. 78). Beyond that line the climate has quite an Arctic character, with few traces of life, either vegetable or animal. Within the torrid zone the mountain contrasts are still more marked. The sultry plains of the Ganges, for example, show a luxuriant tropical vegetation; on the middle slopes of the Himalaya the climates and the vegetation resemble those characteristic of the temperate zone, while the higher summits are like polar latitudes in their temperature and partly in their plants.

9. Every mountain-chain is traversed by a system of **valleys**. Those which run along the line of the chain and separate its ridges, or ranges, are called **longitudinal**; those which run across the ridges and divide them into independent mountains are termed **transverse**. Thus the Swiss Alps contain the two parallel lines of the Bernese Oberland and the Pennine range, separated by the longitudinal valleys of the Rhone and Rhine. The former of these ranges is cut through by many transverse valleys and passes, such as the Rawyl, Gemmi, and Grimsel, between which rise massive snowy mountains. The latter is likewise deeply trenched by valleys and passes, separating some of the monarchs of the Swiss mountains—Mont Blanc, the Matterhorn, Monte Rosa, and the St. Gothard.

10. The direction of the great mountain-chains of the globe has evidently a close connection with that of the continents. Of this connection America, which in many respects is a typical continent, furnishes an admirable illustration. The long, continuous line of mountain-chain that extends from the southern spur of the Andes to the most northerly swell of the Rocky Mountains, a distance of some 9000 English miles, coincides with the

general trend of the continent, and forms the axis or back-bone of that vast tract of land. This main line of elevation in the New World runs in a general north-west and south-east direction. In the Old World a vast mountain-system, with many branches, stretches from the north-east of Asia across the centre of that continent and the south of Europe to the north-western headlands of Spain—a distance of about 12,000 miles. Here the general trend is on the whole east and west. These are the two great mountain ridges of the world.

11. Reserving the structure and origin of mountains for a later Lesson (Lesson XXIX.), after the nature and arrangement of the materials of the solid earth have been examined, we may for the present consider the other leading forms of terrestrial outline already referred to—plains and table-lands.

12. **Plains.**—The chief flat lands of the globe lie between parallel mountain ridges; to a smaller extent they occur as narrow belts or fringes, flanking the sea-ward slopes of mountains or uplands. Take again the American continent by way of illustration. In North America, a vast plain stretches between the Rocky Mountains on the west, and the Alleghany and White Mountains on the east from the Gulf of Mexico up to the Arctic Ocean. Of course this wide expanse of land is not a mere dead level. It abounds with ridges of low hills and lines of valley, but over many thousands of square miles neither the heights nor the hollows are sufficiently marked to destroy its general level character. Westward it slowly ascends until its surface reaches a level of from 4000 to 5000 feet above the sea, and forms a vast table-land from which rise the Rocky Mountains and other parallel ranges farther west. The enormous plains watered by the Mississippi and its tributary rivers are generally fertile grass-covered **prairies**. So flat are they that vessels can sail into them up the rivers for a distance of about 4000 miles from the sea.

13. More than half the surface of Europe is a vast plain bounded by the heights of Scandinavia, Scotland, and Wales on the north-west, by the ranges of the Ural mountains on the east, and by the chains of the Pyrenees, Alps, Carpathians, and Caucasus on the south. From the west of England the plain sweeps eastward across the north of France and Germany, spreads out over most of Russia, and sinking into the depression of the Caspian Sea, passes into Asia, where it bends northward between the Ural and Central Asian mountains, covering there a belt of territory 1000 to 1500 miles broad, and nearly 4000 miles long. In the " Black Lands " of the Russian plain, the soil is a dark rich fertile loam yielding abundant corn-crops. Farther south, where it sinks below the sea-level towards the Caspian, it becomes full of salt and brackish lakes, and assumes a bare and barren aspect. Round the Caspian Sea and northward into Asia the clayey or sandy soil is covered with grass in spring, but is dried up and withered by the scorching summer, and frozen by the severe winter. These tracts are the **Steppes** of Russia. Within 400 or 500 miles of the Arctic Ocean trees and grass disappear from their surface and are replaced by moss; the soil is frozen to a considerable depth, and only thawed for a few inches down in the height of summer. These northern wastes are the **Tundras** of Siberia.

14. Some plains are so placed as to remain mere bare sandy wastes, without vegetation, and incapable of ever being inhabited. The north of Africa, for example, presents to us the widest barren region or **Desert** on the surface of the globe, that of the Sahara. This is a long broad plain, rising here and there into hills and tablelands; but at its eastern end actually sinking to a depth of from 100 to 150 feet below the sea-level. Where scattered springs come to the surface, little spots of vegetation called "*Oases*" make their appearance, but with these sparse exceptions, the vast plain is a region of dry barren shifting sand, where no rain falls, and where the sun's heat is more intensely felt than anywhere else in the world.

15. In the Sahara Desert, as well as in many other plains both in the Old World and in the New, traces of shells are found belonging to species which must have lived in the sea. It can be shown that they have not been transported from the sea to where they now occur, but have lived and died there. Hence they serve to prove that these wide flat or undulating tracts of country must at one time have lain below the waters of the sea. This is an interesting fact which will be afterwards referred to, in its bearings upon the origin of the present contours of the surface of the land.

16. The tracts of comparatively level land are not always low plains. Sometimes they lie at a considerable elevation. When their height exceeds about 1000 feet above the level of the sea, they are usually called **table-lands** or **plateaux**. But, as in the case of the country west of the Mississippi, no hard line can be drawn between plains and table-lands. They often merge into each other. A table-land is only an elevated plain, and, like a plain, it undulates into heights and hollows, is sometimes deeply trenched by valleys and river-gorges, and here and there rises into lofty ranges of hills.

17. A table-land may be enclosed between parallel ranges of mountains. Of this arrangement the most stupendous example is the vast plateau of Central Asia, 10,000 feet above the sea, with its mountain barriers of the Thian Shan on the north, and the Kuen Lun and Himalaya on the south. Another remarkable instance is the vast plateau in Western America, 4000 to 5000 feet high, from which the Rocky Mountains rise, and which stretches westward to the Sierra Nevada and Pacific Coast ranges. It is diversified with numerous parallel mountain ranges, which, like the Rocky Mountains, have a general northerly trend. Its waters descend

each side of the continent, and it also includes the "Great Basin," which has no outlet, and where the lowest hollows are occupied by salt and bitter lakes.

18. Other table-lands rise out of the sea or from low plains bordering the sea. The African continent, for example, is one vast table-land with but a narrow fringe of low plain skirting its shores. Spain and Portugal form another example of a table-land rising steeply from the sea. This well-marked peninsula consists of a plateau about 2600 feet high, crossed by five hill-ranges (in Spanish *Sierras*) and trenched more or less deeply by river-valleys—the Douro, Ebro, Tagus, Guadalquiver, and others.

19. Here and there a table-land has been so extensively worn down into valleys that its former character becomes almost lost. Scandinavia is an illustration of this result. Central parts of the table-land still remain—a vast undulating plateau between 4000 and 5000 feet high. But the outer portions are so traversed by lines of valley as to be cut into narrow ridges. The Highlands of Scotland have been still more thoroughly trenched by valleys until the original table-land can scarcely be any longer traced (Lesson XXIX.).

20. Having noted now the leading forms of surface by which the land is marked, we are led to ask whether any explanation can be given of them. How have the vast mountain-chains been formed? Have they existed from the beginning of things, and have their crests and pinnacles, their precipices, ravines, and valleys, always been as we see them now? An answer to these questions can be given with considerable confidence and fulness, but to be able to understand it we must attend with some care to two aspects of the physical geography of our globe. 1st. The nature of the earth's interior and the reaction of the interior upon the surface; and 2d, the circulation of water upon the land and the other external processes by which the surface of the land is

affected. The next eight Lessons will be devoted to these two subjects, and we shall then return, in Lesson XXIX., to the consideration of the origin of the present features of the land.

LESSON XXI.—*The Composition of the Earth.*

1. Of what materials does the solid land consist? How are these materials arranged so as to form the vast mass of the globe? Man cannot pierce beyond the mere outer skin of the planet; can he then ever hope to learn the probable constitution of its interior? These are some of the questions to which we must now turn. In endeavouring to find answers for them let us begin with what is nearest and most familiar.

2. No matter in what part of the world we may live, the uppermost layer or covering of the land is almost always one of vegetation; in one place grass, in another forest, in another thickets of jungle or brushwood. Here and there, indeed, the surface is formed of shifting sand, where no plants may be able to take root, or of rough mountainous ground, where ledges and crags of bare rock shoot up into the air. But even in such places as these straggling weeds or shrubs may be found striving to bind the loose sand together, or to find a lodgment among the crannies of the rock.

3. Beneath the cover of vegetation we see the layer of soil on which the plants grow, sending their roots down into it and extracting from it the soluble mineral materials which are needed for their framework. Soil varies greatly in colour and composition, being sometimes a stiff, gray clay, sometimes a soft, black, crumbling loam, sometimes a brown or yellow sand, and sometimes so stony as to be little else than a sheet of gravel. In all cases, however, it will be found to consist of small particles or larger fragments, which

appear broken or worn, and have been evidently derived from some solid rocks. Looked at with a magnifying-glass this broken and derivative character appears still more clearly. Roots of plants descending through the soil open it up and permit rain and air to pass into it, while the common earth-worm likewise aids in this work by swallowing the fine parts of the loam, voiding them above ground, and thus bringing up lower portions of the soil to the surface.

4. Besides the grains of sand, earth, or clay, the soil

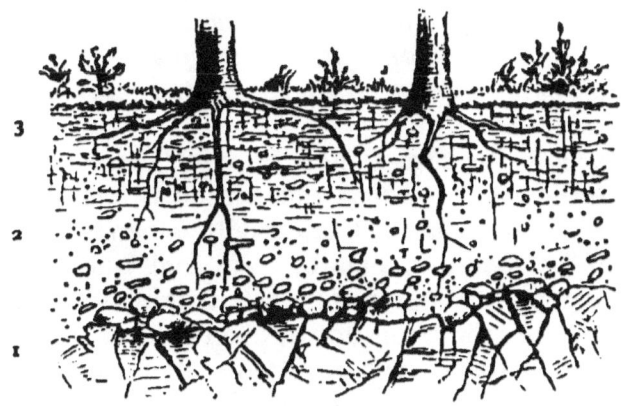

Fig. 25.—Section to show the formation of (3) soil and (2) subsoil from rotting away of (1) rock underneath.

contains more or less organic matter derived from the mouldering remains of plants and animals. This ingredient is essential to the fertility of the soil, and is made use of by living plants. Where arable land has been cultivated for a long time, each successive crop has taken so much soluble mineral materials and organic matter out of the ground that the soil becomes exhausted, and needs to be renovated by the introduction of various kinds of manures before it will continue to yield good crops. In some countries, where the soil is deep and fertile, many centuries may elapse before such exhaustion takes place.

5. The soil varies greatly in depth. A common thick-

ness is three or four feet, but on sheets of rock it may not be so much as an inch, while in rich plains it may be several yards. The next layer below it is called the **subsoil**. It consists of the same materials as the soil, but less finely broken up, and with little organic matter. Indeed, the soil is simply the upper part of the subsoil which has crumbled away and been mixed up with the decayed remains of plants and animals. Usually only the longer roots of the larger plants, such as forest-trees, reach down into the subsoil. But as the soil is washed off by rain from the surface of the ground, the upper parts of the subsoil, coming in consequence nearer to the surface, are more exposed to air, percolating rain, plant-roots, and earth-worms, so that they gradually pass into soil.

6. Beneath the subsoil lies the **rock**, from the breaking up and decay of which the subsoil is formed. The character of the soil thus essentially depends upon the nature of the rock underneath. We shall trace in subsequent Lessons how universally and strikingly the surface of the land is crumbling down, even when it consists of the most solid rocks, such as granite, limestone, or sandstone. For the present we may bear in mind that the covering of soil, so generally distributed over the land, and on which the verdure and fertility of a country depend, has been formed by the gradual decay of the rocks and of the remains of the successive generations of plants and animals that have lived upon the surface.

7. Beneath the decaying soil and subsoil, then, we arrive at the undecomposed rock. In most countries of the world a great variety of rocks occurs. To describe these and to trace their origin and history forms the subject of the science of Geology. But a little consideration may here be given to some of their more important characters, as it will enable us to understand better how the materials of the land have been put together, and what is the probable constitution of the interior of the planet.

XXI.] THE COMPOSITION OF THE EARTH. 185

8. In the first place, then, the most cursory observation suffices to show that most of the rocks of the land have, like the soil, been formed out of worn fragments or particles of older rocks. Sandstone, for instance, out of which so much of the solid framework of the plains, hills, and mountains has been built, consists of mere sand compacted into solid stone. Shale and slate are only so much hardened clay or mud. Conglomerate or puddingstone is nothing but a mass of solidified gravel. In all these cases the materials of which the rocks are composed have been worn away from still older rocks, and have been rolled to and fro in water, as gravel, sand, and mud are produced at the present day.

FIG. 26.—Bedded arrangement of rocks. (*a*) conglomerate; (*b*) sandstone; (*c*) shale.

9. Rocks formed of these water-worn materials have been accumulated to depths of many thousand feet. They now frequently form lofty ridges of mountains. They spread also over the lower grounds, and form the wide plains of the world. One of their most obvious features is their arrangement in layers, beds, or strata, which, varying in thickness from less than an inch to several feet or yards, are piled over each other. Hence a cliff of these rocks has a striped or banded look

(Fig. 26). They are known as **bedded** or **stratified** rocks.

10. Together with these there often occur other rocks formed wholly or in great measure of the remains of plants or animals. For instance, amid a mass of sandstone and hardened clay, traces of the fronds of ferns (Fig. 27), and the seed-cones, leaves, stems, and roots of other plants, may occasionally be observed. These vegetable remains have in some places been so crowded together

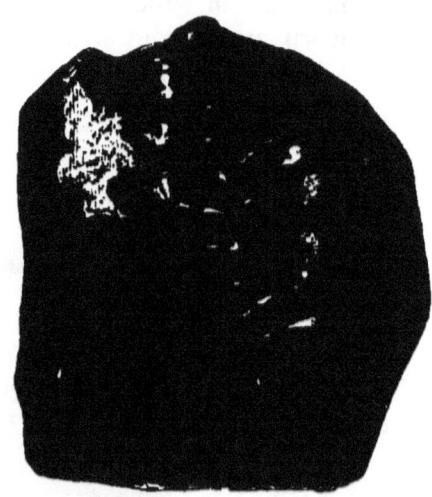

FIG. 27.—Piece of shale containing portion of a fossil fern.

as to form black or brown seams of **coal**, and in that condition supply much of the fuel which man consumes at the present day.

11. Again, many rocks are composed mainly or entirely of broken shells, corals, and other animal remains. Limestones, for example, commonly have this composition, as shown in Fig. 28, which represents a piece of limestone made up of the crowded joints of the encrinite or stone-lily—a marine animal of which there are some small modern kinds still living. Chalk is a variety of limestone consisting mainly of the broken remains of

XXI.] THE COMPOSITION OF THE EARTH. 187

minute forms of marine animal life (Fig. 29) like those that compose the modern ooze of the Atlantic sea-bottom (Fig. 14).

12. Limestone, made up entirely of the crowded remains of corals, shells, sea-urchins, and other sea-creatures, is extensively formed at the present time upon submerged banks in the tropical seas. Rock of a similar kind, and anciently formed in like manner on the sea-

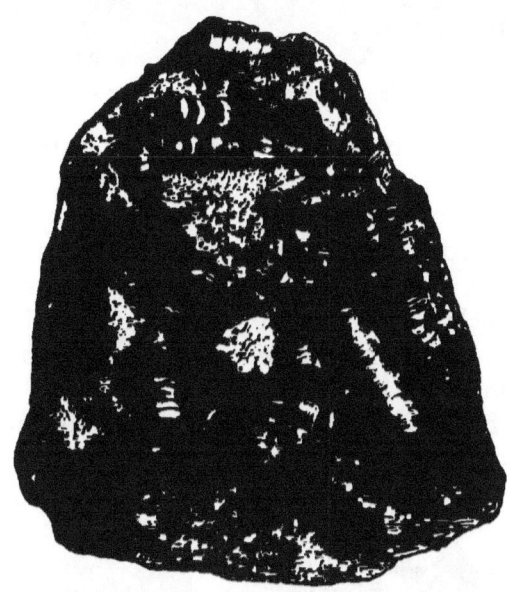

FIG. 28.—Piece of limestone, showing how the stone is made up of animal remains.

floor, covers thousands of square miles in many countries, and not only underlies parts of the plains, but, like sandstones and shales, rises up even into lofty mountains. The outer ridges of the Alps, Himalaya, Rocky Mountains, and Andes, for example, are in large measure built up of limestones and other rocks, formed mainly or in part of animal remains.

13. It appears, then, that most of the land, as far as we can see it, is made of rocks, which are either hard-

ened gravel, sand, and clay, or have been formed out of the broken and compressed remains of once living plants and animals. From these facts we conclude first, that

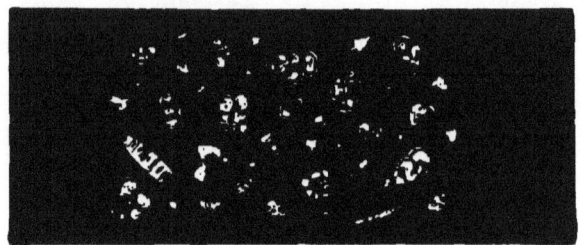

Fig. 29 —Some of the grains of a piece of chalk.

what is now land must have been under water; secondly, that as the limestones and other strata contain chiefly marine forms of life, the water in which most of them were deposited must have been the sea; thirdly, that the materials out of which the land has mainly been formed were laid down in the sea; and fourthly, that by some means these ancient deposits of the deep have been upheaved into dry land.

14. In the second place, besides the various stratified rocks, which have been chiefly derived from the wearing

Fig. 30.—A piece of granite, showing the composition of a Crystalline Rock.

down of rocks older than themselves, others may be seen in which no bedded or stratified arrangement exists. These are not made up of fragments derived from previously-formed rocks, but, for the most part, are **crystal-**

line, that is, are made up of crystals, either felted together or imbedded in a glass or base. Granite, porphyry, and basalt are examples of such crystalline rocks. Instead of spreading out in vast sheets over the continents, as the stratified series does, the crystalline rocks occur in bands or bosses, and rise through the stratified rocks sometimes in huge disruptive masses, sometimes in veins and variously-shaped intrusions. They are particularly to be noticed along the central portions of mountain-chains, where they may be seen coming up through the very oldest of the stratified rocks. They likewise appear in the eruptions of volcanoes—the molten rock known as lava being one form of the crystalline rocks.

15. There can be little doubt that, as a rule, the rocks of the crystalline series have come from below, and have been thrust in a melted state among the other rocks, or have been poured out at the surface as lava. Hence we must infer that beneath the outer layer of stratified or derivative rocks, though it may be many thousands of feet in thickness, there must be an inner layer or mass of crystalline material, which has here and there been squeezed through the stratified rocks, as in the axis of mountain-chains, or has communicated with the surface by means of the openings of volcanoes.

16. So far these conclusions, being founded on what can be actually seen, may be regarded in the light of established truths. We have traced the materials of the solid earth from the thin outer layer of surface-soil down through the thick piles of stratified sandstones, clays, limestones, and other rocks into the still lower granite, lava, and other crystalline masses. We have seen that most of the materials of the land have been raised up out of the sea, and that portions of the underlying crystalline rocks have actually been pushed up into the very heart of the mountains. Can we descend any farther and trace still deeper layers in the structure of our planet? Not directly; no distinct lower portions have

come through the crystalline rocks. Nevertheless, evidence enough may be gathered to indicate with some probability what lies still farther down.

17. Most of the rocks at the earth's surface weigh from twice to thrice as much as water. In other words, their specific gravity is from 2·0 to 3·0, that of pure water being reckoned as 1·0. Experiments made with the pendulum and the plumb-line on the earth's attraction, indicate that the weight of the planet is about twice as much as that of the rocks at the surface, or, more shortly, the density of the earth is about 5·5. Obviously the interior of the earth consists of materials heavier than those that form the land.

18. Owing to the increase of the force of gravity, the density of every substance becomes greater in proportion as it approaches the centre of the earth. But we know very little of the rate of compressibility in the high temperature that exists within the earth. If the increase of density continued unchecked, air would be as heavy as water at a depth of 34 miles, and water would be as heavy as mercury at a depth of 362 miles. But the effect of heat is to expand bodies and thus to diminish their density, and doubtless the earth's internal heat prevents the density of the whole planet from being so great as it would otherwise be.

19. The deepest mines do not reach one mile in depth, or $\frac{1}{3982}$ of the distance from the surface to the centre of the earth. Man can never, indeed, hope to penetrate far into the interior of the planet on which he lives. There are yet several kinds of evidence open to him which go to demonstrate our globe's high internal temperature. These are: 1st, mines, wells, and borings; 2d, hot-springs; and 3d, volcanoes.

20. (i.) It has long been known that the air of deep mines is warmer than that above ground, and that, as a rule, the deeper the mine the warmer is the air in it. A coal-mine sunk at Rose Bridge near Wigan, for

example, to the great depth of 2445 feet has a temperature at the bottom of 94° Fahr., while the temperature at the surface is only 50° Fahr., which is a mean increase of 1° Fahr. for every 55 feet of descent. But the rate varies greatly at different places and even at different levels at the same place. The water that rises from deep borings is warm. Thus from a well sunk to a depth of 1798 feet at Grenelle, near Paris, water issues with a temperature of 81·7° Fahr. As the results of observations made all over the globe it appears that an increase of temperature always occurs, and that though it varies in amount in different kinds of rock, its average rate is about 1° Fahr. for every 60 feet of descent.

21. If this rate should continue, it is plain that at a comparatively short depth even the most refractory substance would be at its melting-point, though the effect of the enormous pressure within the earth may prevent it from actually passing into a molten condition. Water at 12,000 feet would be as hot as boiling water at the surface. At a depth of about 24 miles, if its component gases could remain united, it would be as hot as melting gold.

22. (ii.) In almost all countries of the world springs of warm water rise from the ground. In volcanic districts the water is often even at the boiling-point, and remains permanently so. Warm springs often occur, however, at a distance from any active volcano. Those of Bath, for example, rise with a temperature of 120° Fahr., in the lowlands of the south-west of England, more than 1000 miles from the burning mountains of Iceland on the one hand and more than 1100 from Vesuvius on the other. At Buxton in Derbyshire the springs have a temperature of 82°. In Germany the wells of Wiesbaden throw out water at a temperature of 158°, and those of Carlsbad at 167°, while in the north-west of Spain some of the springs have a temperature as high as even 192°. All these places are remote from an active volcano, but it may often be

noticed that hot springs rise along mountain-chains, or at least on lines where the rocks have been intensely crumpled and where they may have been greatly heated during the crumpling. Here and there, as will be pointed out in the next Lesson, the remains of volcanoes occur which are now quite cold and silent, and from which no eruption has taken place within the memory of man. Yet some of these extinct volcanoes are still associated with hot springs. In the ancient volcanic district of Central France, for example, numerous springs occur, sometimes with a temperature of as much as 174°.

23. By far the most remarkable kinds of hot springs, as showing the great heat of at least some parts of the earth's interior, are those which throw out their contents with violence. They occur in volcanic districts, and are called **geysers** (that is, gushers), that being the local name borne by those of Iceland, which were first described. The part of Iceland where they chiefly occur is a space about two miles square in a low, wide valley among volcanic rocks. In this limited tract the ground is pierced with abundant openings from which jets of steam and hot water escape. These openings, said to be nearly 100 in number, are usually surrounded with basin-shaped rims or mounds of white, gray, and variously coloured incrustations of silica called *sinter*, deposited from solution in the hot water. They vary in size from mere tiny bowls, a few inches across, up to huge cauldrons, like the Great Geyser, which rises 15 feet above the ground, and measures 56 feet in diameter. In the centre of the basin of the Great Geyser a pipe or funnel, eight feet wide, descends into the earth. From this opening boiling water is constantly ascending into the basin and flowing over the lowest part of its lip into the plain. At intervals of a few hours a loud rumbling is heard to proceed from the funnel, the water in the basin begins to boil up, and jets of it together with clouds

XXI.] THE COMPOSITION OF THE EARTH. 193

of steam are spurted up for a few feet. After these intermittent eruptions have continued for about a day, they are succeeded by a much more violent explosion. The ground trembles slightly, as the rumbling sound increases in violence, until a huge column of boiling water, 150 or 200 feet high, is thrown up into the air, with

FIG. 31.—View of the geysers of Iceland. Great Geyser in eruption.

loud explosions and clouds of steam. In this way the basin and pipe are emptied of water, but they gradually fill again, the rumblings and jets of water and steam begin anew, until next day another grand outburst empties the geyser. It has been found by observation that the lower parts of the subterranean column of hot water are sometimes actually raised to a temperature of 261° Fahr., or 49 degrees above the common boiling-point. Hence when this superheated water rises to near the surface, and is thereby relieved of the pressure of the column of water that previously lay above it, it flashes into steam with a loud roar, and throws out jets of boiling water and steam.

24. The Yellowstone National Park, in the Territories of the United States, contains several hundreds of geysers scattered over a volcanic tract of country, some exceeding the Great Geyser of Iceland in size and in the height and volume of water and steam which they discharge when in eruption. One of the most remarkable, known as "Old Faithful," emits a lofty jet of hot water and steam, with great regularity, about once an hour. The sites of the underground foci appear to shift their positions from time to time. Some of the vents have evidently been only recently opened, for the trees invaded by their deposits are hardly dead yet. Others are no longer active geysers, but mere quiet pools of hot water, or mounds, whence only steam rises. But many are quite extinct; their rims and basins of siliceous incrustations, standing above the ground, mark where the boiling water once rose, and the surrounding forest is beginning to encroach upon their decaying heaps of sinter.

25. Another interesting series of geysers occurs in New Zealand, where also active and extinct volcanoes are met with. The spouting hot springs of Orakeikorako gush forth in numbers on either side of a river-valley, while at Tetarata the water is so largely charged with silica that it has formed a series of terraces and basins down the precipitous bank from near the top of which it takes its rise.

26. (iii.) **Volcanoes** form so interesting a part of the earth that they will be described in some detail in the following Lesson. We need only note here that being openings from which steam and other hot vapours as well as streams of melted rock are discharged, and occurring in many parts of the world, they prove that within the earth the temperature must be at least as high as the white heat at which lava issues at the surface.

27. It is plain, then, from such evidence that the earth cannot but have an enormously high internal temperature. Much dispute has arisen as to whether the interior is liquid or solid. Some have maintained that the earth consists of a ball of molten material with an exterior crust which has been variously estimated at from 20 to 1000 miles in thickness. Others have insisted that a planet so constituted could not rotate as the earth does, and that our globe must be solid to the centre.

28. If the heat goes on increasing at the same rate as it is observed to do in the comparatively shallow borings which man can make into the earth, the ordinary melting-point of even the most refractory substance would, no doubt, be met with at a comparatively short depth beneath the surface (Art. 21). But it does not follow, therefore, that the materials of the interior should be actually in a state of fusion. Pressure is believed to have the effect of raising the melting-point of such substances as those which form most rocks, that is, it keeps them from melting until a still greater heat is applied. The pressure at great depths within the earth must be enormous. Hence below a depth of a few miles, where the temperature reaches the ordinary melting-point of most rocks at the surface, each successive layer of the earth's substance may not be actually fused, but may be just at the melting-point proper to its depth. The whole globe might thus be solid, but the least diminution of pressure at any point would allow the parts so relieved to melt at once. This, so far as can be judged, seems to be the most probable condition of the interior of the earth.

29. The internal heat being so great, there seems no

reason why the innermost parts of the planet may not consist of metal, such as iron or gold. And indeed there is some reason to infer that they are really metallic. Cracks have been made abundantly through the rocks forming the land, and in many of these occur metallic ores which are believed to have not improbably come up from a metallic region below. Researches into the constitution of the sun and of the other planets have tended to confirm this view of the metallic composition of the central parts of the earth.

30. Summing up what has been said in this and previous Lessons regarding the constitution of the globe, we conclude that beneath the outer envelopes of the atmosphere and the sea, the solid earth has an upper layer of loose, crumbled materials—gravel, sand, and mud on the sea-floor, soil upon the land—under which lies a thick series of stratified and derivative rocks ; that still farther down masses of crystalline rocks, which have in many places been forced through the overlying series, descend to an unknown depth, and that within them there may be a metallic central core. We see, moreover, that since the heat rapidly increases as we descend, the innermost parts of the globe must have a temperature, probably more than sufficient to melt every known substance, but that the pressure of gravity may be great enough to retain the general mass of the globe in a solid state, except at such places as supply the streams of molten lava that flow from volcanoes.

LESSON XXII.—*Volcanoes.*

1. The word "volcano" is derived from Vulcanus, the name of the Roman god of fire, who was supposed to have his subterranean forges at the roots of the mountain Etna. It is applied to any conical mound, hill, or mountain, formed of materials which have been erupted

in a molten or solid condition from beneath the surface. A volcano is **active** when it ejects with violence gases, steam, water, mud, dust, stones, or molten rock from its summit, or from fissures or other openings on its sides. The term **volcanic action** is used to describe all the kind of work done by a volcano. A volcano may be **dormant** when it remains a long time without any eruption; and it is said to be **extinct** when, though its external form may remain, it does not seem to have been in eruption for a great many centuries, and gives none of the usual signs of volcanic action.

2. The size of a volcano varies from a mere little mound, a few yards in diameter, like some of the mud-volcanoes around the Caspian Sea, up to a giant mountain like Cotopaxi, which rises among the Andes to a height of 18,887 feet above the sea, its upper 4000 feet forming a smooth snow-covered cone, with an orifice at the top, whence hot ashes and stones are scattered far and wide over the surrounding country.

3. At the top of a volcano lies a basin-shaped hollow called the **crater**, from the bottom of which the pipe or shaft descends, whereby the volcanic products are brought to the surface. Showers of fine dust and stones are frequently thrown out from volcanic craters. These materials falling down the slopes of a cone gradually increase its diameter and height. In like manner streams of molten rock called **lava**, issuing either from the lowest part of the lip of a crater or from some fissure or orifice on the outer declivity, pour down the slope and harden there, thus still further augmenting the bulk of the hill.

4. As a volcano increases in size, and cracks are formed in weak parts of the cone, smaller cones are piled up on its flanks by the emission of dust, stones, and lava from these fissures. A large volcanic mountain, like Etna, or the Peak of Teneriffe, is thus sometimes loaded with small volcanoes which may reach a

height of five or six hundred feet. In the drawing (Fig. 32) a plan is shown of one of these great volcanoes with its groups of minor cones.

5. At the beginning of a volcanic eruption rumblings are heard, like the muttering of distant thunder, while the ground is felt to tremble slightly. These noises and

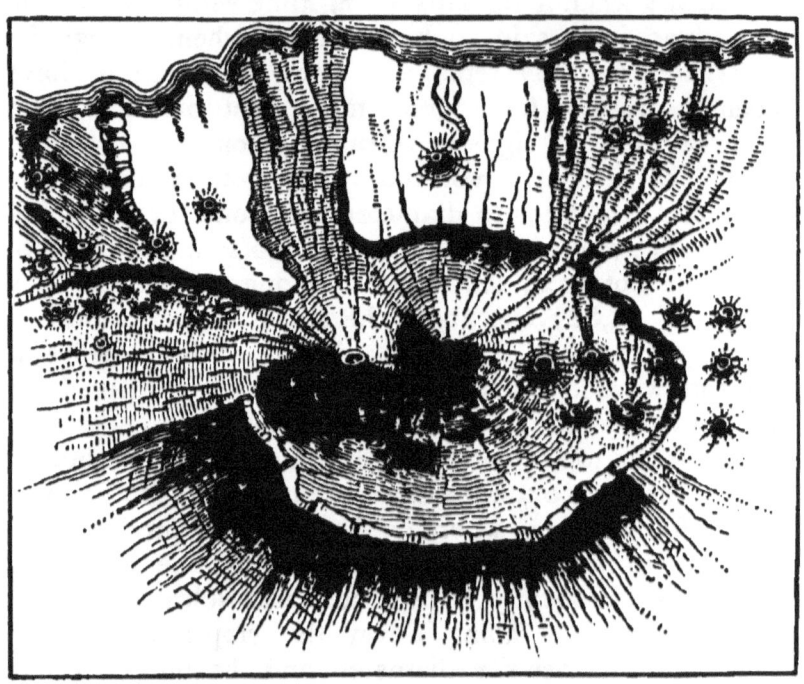

Fig. 32.—Plan of the peak of Teneriffe, showing the large crater, partly effaced, and smaller craters with lava currents issuing from them.

tremors increase in intensity, successive loud explosions take place in the pipe of the volcano, and at last clouds of fine dust and steam are hurled with prodigious force far up into the air. The steam rapidly condenses into rain, which falls in torrents down the outer slopes of the mountain. The fine dust is sometimes given out in such quantities as to darken the sky for many miles around. In the famous eruption of Vesuvius, which in the year

79 destroyed the Roman cities, Herculaneum, Pompeii, and Stabiæ, the air was as dark as midnight for twelve or fifteen miles round, and a thick deposit of fine ashes and stones fell on the whole district. Some notion of how completely this eruption of ashes covered the country may be obtained from the accompanying drawing, which

FIG. 33.—View of a Street in Pompeii.

represents one of the streets of Pompeii as it now appears, after the volcanic deposit under which the city lay buried for sixteen centuries has been cleared off. At the farther end of the street where the excavations have stopped, the dark layer on the top of the walls, and on which the pine-tree is growing, represents the thickness of the general covering of volcanic ashes. In some volcanic eruptions the finer dust, carried up into a strong upper current of air, has been known to be transported for hundreds of miles before descending to the ground. Illustrations of this kind of transport were cited in Lesson XI. to prove the existence of these upper currents in the atmosphere.

6. Besides the light dust, vast numbers of white-hot or red-hot blocks and smaller fragments of stone are ejected from the crater. Many of these strike against each other as they rise and fall, producing at night a wonderful spectacle as their sparks and flashes light up the darkness. Of the force with which the stones are occasionally discharged some notion may be formed from the feats of the volcano Cotopaxi, which is said to have thrown a block, estimated to weigh 200 tons, to a distance of nine miles. The volcano of Antuco in Chili is reported to have sent stones flying to a distance of thirty-six miles.

7. It is not difficult to understand why, in the earlier stages of a violent eruption, so vast an amount of fragmentary materials should be ejected. During the successive explosions, the walls of the crater and the accumulated lava and other volcanic ejections, which had gradually choked up the funnel, are rent and finally blown into fragments. It is possible that at the deep roots of a volcano the temperature is so high as to keep the elements of water dissociated. Where these elements can still remain combined, under enormous pressure, and at a temperature far above the boiling-point, the water, which has perhaps saturated the molten and solid rocks in the pipe of the volcano, instantly explodes, when, in its upward ascent, it can overcome the influence of the superincumbent pressure. So violent is the conversion of such superheated water into steam that the molten lava is actually blown into the finest dust. It is the liberation of successive portions of this highly-heated water into the gaseous state that produces the explosions which form so magnificent a part of a great volcanic

Equator
Sumatra
Zanzibar
Seychelles
Chagos I.
INDIAN OCEAN
Java
Madagascar
Mauritius
Bourbon I.
Tropic of Capricorn
N.W. Cape
AUSTRALIA
Port Natal
C. Leeuwin
Amsterdam
St. Paul
Crozet I.
Kerguelen Land
Tasmania
TIC OCEAN

eruption. At each explosion a vast ball of steam shoots up into the air, at once condensing into white clouds and rolling over in huge folds (Fig. 34), which either dissolve and float away in the higher atmosphere or are further condensed and fall as rain.

8. The upper part of a cone, during the early part of

Fig. 34.—View of Vesuvius as seen from Naples during the eruption of 1872.

a tremendous eruption, sometimes disappears, because it is shattered by the explosions and blown up into fragments, which fall back either into the crater or down the outer slopes of the mountain. Mount Vesuvius has supplied some excellent illustrations of this kind of destruction. Prior to the first century of the Christian era that mountain was a dormant volcano, from which no explosions had ever been known to come, but which had an enormous crater on its summit, overgrown then with brushwood and wild vines, like the crater of Astroni and some of the other extinct volcanoes near Naples (see Fig. 5). Suddenly in the year 79, when the great eruption took place which destroyed Pompeii, the southwestern side of the cone was blown away and a new cone of much smaller dimensions was formed inside the circuit of the former crater. In the drawing of Vesu-

FIG. 35.—Mount Vesuvius as seen from the sea, with the remaining part of the old crater of Somma behind.

vius (Fig. 35) as seen from the sea, the crescent-shaped half of the old crater appears behind the modern diminished cone. In August 1883 the volcanic cone forming the island of Krakatau in the Sunda Strait was almost entirely destroyed by one of the most colossal volcanic explosions of modern times.

9. The enormous expansive force of imprisoned water and steam drives the molten lava up the pipe of the volcano. After the first explosions, during which a column of molten rock has been propelled upwards in

the funnel, lava is seen to flow either from the top or from one or more points on the side of the cone. Should the sides of the mountain be solid enough to resist the enormous pressure of the ascending lava, the latter will, of course, find no escape until it fills up the crater to the level of the lowest part of its rim, over which it will pour down the mountain. More usually, however, there are weak parts, such as rents, caused by the previous explosions, through which the lava finds egress to the outer slopes of the cone.

10. Few sights in nature are more terrible than that of a lava torrent as it pours down the side of a mountain. At first, glowing with a white light, it flows freely like melted iron; but rapidly grows red and darkens. Its surface soon hardens into a black or brown crust that breaks up into rough cinder-like pieces, beneath which, as may be seen at the rents, the main mass still remains red-hot. A short way from the point of emission, the lava-stream looks like a river of rugged blocks of slag, grinding against and over each other, with a harsh metallic sound, and revealing, every here and there, a glimpse of the fiery flood underneath, over which they are floating. Clouds of steam and hot vapours rise from all parts of the moving mass. Its rate of march varies much according to the slope of the ground, distance from the point of exit, and other causes. Thus, in the year 1805, a current of lava ran down the first three miles of the slope of Vesuvius in four minutes, yet took three hours to reach its farthest point, which was only six miles. In 1840 one of the more liquid lavas of Mauna Loa, in the Sandwich Islands, was observed to run eighteen miles in two hours.

11. The front of a slowly advancing lava-current seems to be a huge mound of rubbish, somewhat resembling one of the piles of debris at ironworks, or a new railway embankment, with its raw earth and stones. The rugged blocks of lava tumble over each other as the mound

slowly advances. A breeze coming from the lava is hot and stifling, for it carries with it some of the steam and acid vapours that stream out abundantly from the molten rock. Trees, walls, gardens, and vineyards are successively buried under the burning flood. A house or range of houses may form a temporary barrier; but

Fig. 36.—Houses surrounded and partly demolished by the lava of Vesuvius, 1872.

eventually these buildings are pushed over like mere houses of card and engulfed (Fig. 36), or the lava is piled up against, and passes round them, meeting again below so as to leave their roofs, perhaps, projecting from the middle of a rugged sea of lava. It occasionally happens that the still liquid rock inside bursts through the hardened crust at the side or in front, and pours down in a new direction. This is always one of the risks to be considered in such populous and cultivated tracts as those along the slopes of Vesuvius.

12. The distance to which a torrent of lava may flow can never be foretold, even when it is seen pouring down from a mountain at a great velocity. At Vesuvius many of the lava-streams have reached the sea; others have not travelled farther than a few hundred yards from their point of emission. The most tremendous floods of lava ever known to proceed from a volcano were those of Skaptar Jokul in Iceland, in the years 1783-5. Vast torrents of molten rock were then poured over that island, filling up river-courses, ravines, and lakes, and completely destroying the surface for many hundreds of square miles. The lava reached to an extreme distance of forty-five miles in one direction, and of fifty miles in another. In some level places, where the lava spread out, the stream reached a breadth of fifteen miles, and a thickness of 100 feet, but it accumulated in narrow valleys sometimes to a depth of 600 feet. It has been computed that the total mass of lava poured forth during that series of eruptions would form a mountain equal in bulk to Mont Blanc.

13. The hardened crust on a lava-stream forms an excellent non-conducting layer between the still molten mass underneath and the air. One can walk on that crust, though, through the fissures, the red, glowing lava may be seen to lie only a few inches below. It is evident that the inner parts of such a current, even at a considerable distance from the point of emission, may retain a very high temperature, long after the current itself has come to rest. Thus the lava of the eruption of Vesuvius in 1858 continued to give out steam and hot vapours in 1873, and still retained so much heat that one could not keep one's hand in some of the fissures for more than a few seconds, although the surface of the current was everywhere quite cold.

14. When the lava begins to flow freely from a volcano, the violence of the eruption usually abates. The showers of ashes gradually cease, or at least do not extend beyond the crater and its immediate vicinity. The explosions and subterranean rumblings disappear, and except for the cloud which always hangs over the summit of the mountain and marks how constantly and abundantly steam still continues to rise from the crater, nothing at last remains to indicate from a distance that volcanic action is not extinguished, but merely quiescent for a time. Yet a visit to the summit of the cone at such an interval would show that hot vapours and gases still keep streaming out from the summit and sides of the mountain. Occasionally after an eruption of Vesuvius it is observed that a large destruction takes place among hares and birds on the hill-sides, owing to the plentiful emanation of the poisonous carbonic acid gas. Many centuries after a volcanic district has ceased to be subject to any eruptions this gas continues to rise, sometimes in large quantities, either bubbling through the water of springs or coming out of crevices in the ground.

15. The active volcanoes on the surface of the globe are not distributed at random, but follow certain lines, as shown in the map on Plate IX. It will be observed that these lines generally coincide with some of the ridges on the surface of the earth already referred to (Lesson XIX.). Almost all volcanoes are ranged close to one or other of the great ocean basins. The continental and island barriers which encircle the Pacific Ocean form one vast ring of active volcanoes. Beginning on the east side we find in the giant chain of the Andes a series of active volcanoes, some of them the loftiest on the globe, running along the western margin of South America. This series is continued through Guatemala and Mexico into North America, and stretches through the Aleutian and Kurile Islands, forming a close-set fringe to the northern Pacific Ocean. The line is

prolonged down Japan, the Formosa and Philippine Islands, to the Malay Archipelago, where it divides into two branches. One of these, turning south-eastward by New Guinea and the New Hebrides into New Zealand, is prolonged, but with wide gaps, across the Pacific basin in the volcanoes of the Friendly, Society, Marquesas, and Easter Islands, towards the coast-line of South America, thus completing the vast volcanic ring. Even along the centre of that basin on the submarine ridge alluded to (Lesson XII. Art. 18), magnificent active volcanoes appear in the Sandwich Islands, some of those in Hawaii exceeding 13,000 feet in height.

16. Returning to the Malay Archipelago we observe that the second branch of the volcanic line turns north-westward through Java and Sumatra, where more active and dormant volcanoes are crowded into a shorter space than anywhere else in the world, and where the eruptions have sometimes been on a colossal scale (Art. 8). It is prolonged through the islands off the west coast of Burmah. After a wide interval it reappears in Mantchouria, then on the southern borders of the Caspian Sea, whence it may be traced by the Greek Archipelago, Vesuvius, and the Italian Islands, to the Azores, Canaries, and Cape Verde Islands.

17. Besides these main lines, however, scattered volcanoes occur on other distant islands or on the edges of the continents. In the Arctic Ocean lie the active craters of Iceland and Jan Mayen. On the west side of the basin of the Indian Ocean we find small volcanoes on the Red Sea and in the Isle of Bourbon. The line of volcanoes in Terra del Fuego at the southern apex of America seems to be continued in the chain of the South Shetland Islands, and to extend even into the supposed Antarctic continent, where, amid the vast snow-fields of that region, Sir James Ross in 1841 discovered an active volcanic cone 12,369 feet high.

18. But besides active volcanoes there is a still larger

208　　　PHYSICAL GEOGRAPHY.　　[LESS.

FIG. 37.—View of the extinct volcanoes of Central France, taken from the Puy de Pariou.

number which are either dormant or extinct. It would seem that few large tracts of land exist where evidence may not be obtained of former volcanic action. Lava-streams and consolidated beds of volcanic dust may be found in almost all countries. Sometimes, indeed, as in Central France (Fig. 37) the cones are still as fresh as if they had been thrown up only recently; and yet no record remains that they have ever been in eruption within the times of human history. Hence, if to the present long list of active volcanoes be added those which are now extinct, the whole surface of the land will be found to be studded over with points of volcanic eruption.

19. When we consider that each of these points marks an orifice at which highly-heated materials are emitted now, or at least have been emitted at some former period, and that they are so widely distributed over the earth's surface, we see how important is their evidence as to the high internal temperature of the earth (Lesson XXI. Art. 27).

20. It was pointed out (Lesson XXI. Art. 29), that while the interior of the earth is probably solid as a whole, each portion of it, beyond a depth of a few miles, is probably at the melting-point and ready to pass into a liquid condition when any diminution of the pressure takes place. The ridges on the surface of the earth, formed, as they have been, during the contraction and consequent general subsidence of the outer parts of the planet, have doubtless by their uprise relieved the pressure upon the parts underneath them. This relief has probably allowed portions of the interior to pass into the state of fusion. Observe how the active volcanoes of the globe are mostly arranged along such lines of elevation, whether on a continent, as in South America, or in chains of islands, as on the western and northern sides of the North Pacific Ocean. This arrangement can hardly be accidental. It helps to connect the elevation of the

land and the phenomena of volcanoes by showing that long spaces of melted rock should be expected to lie under those very regions where active volcanoes occur. Volcanoes do not pierce every mountain-chain, however, though in many cases they can be shown to have once existed there, but to have been long extinct.

21. Reservoirs of lava may exist underneath without giving rise to actual volcanic explosions, so long as no passage is opened to the surface, and nothing occurs to determine volcanic excitement. The vapours absorbed in lava may be partly original constituents of the earth's interior, partly derived from water which, supplied by rain, rivers, lakes, or the sea, filters through the upper rocks. In the deep, intensely hot regions, whether its gases remain combined or not, the absorbed water must exist at an enormously high temperature, and in a condition that would be impossible save under enormous pressure. Portions of this superheated water that may succeed in effecting their escape and in blowing out an opening in the earth's crust, relieve the pressure on the deep-seated mass of molten rock which is then impelled upwards to the surface. With a loud explosion the liquid rock is blown into dust as its water flashes into steam, and lower portions then begin to flow out tranquilly as streams of lava.

LESSON XXIII.—*Movements of the Land.*

1. Along certain lines, or at certain points, where communication has been opened between the surface and intensely hot materials below, showers of dust and stones, and streams of melted rock, have been emitted in such quantity as to form huge mountains like Etna, the Peak of Teneriffe, and Cotopaxi. But other notable effects show the influence of the internal condition of the globe upon its surface. The solid earth is subject to movements either sudden and violent, or slow and imperceptible. It is sometimes convulsed and rent open, sometimes one tract is gradually raised up above the sea-level, while another is step by step depressed. We shall consider these movements under the three divisions of—1st, Earthquakes; 2d, Upheaval; 3d, Subsidence.

2. (i.) **Earthquakes.**—These, as the word denotes, are tremblings or concussions of the ground. They vary in intensity from mere slight tremors, like that caused by a loaded waggon moving along a street, up to such violent catastrophes as those in which the ground is thrown into undulations whereby rocks are loosened, trees are shaken out of the soil, and villages and cities are levelled to the ground.

3. It is chiefly the terrible devastation of life and property which is usually dwelt upon as the main feature of importance in earthquakes. In a moment, without the least warning of any kind, a rumbling like that of distant thunder or the firing of cannon is heard or felt, and before men have had time to ask each other what the tremor can be, the earthquake comes upon them. The ground rises and falls under their feet; houses rock to and fro until they are rent from top to bottom, or fall with a crash into ruins; the earth here and there opens and closes again. In a few seconds a whole city is demolished, and hundreds or thousands of its inhabitants are dead or dying. The Lisbon earthquke in 1755, for instance, destroyed 60,000 human beings. The Calabrian earthquake a generation later proved fatal to not less than 40,000. Nor is the destruction confined to one limited district. Frequently it spreads over thousands of square miles, carrying death and havoc far and wide through different provinces and states.

4. When, however, we regard earthquakes as part of the machinery of nature, it is not so much their influence upon human life and property as their permanent effects upon the surface of the land which claim our notice.

An earthquake is not felt simultaneously over the whole region affected by it. It begins at one side or end, and travels rapidly to the other, or it is experienced first and most violently over a certain central space, from which it spreads with diminishing intensity in all directions. The sound of its approach, like that of distant thunder, artillery, or rumbling waggons, may precede by a few seconds the advent of the earthquake itself. When the shock comes, the ground is felt to be alternately raised and depressed, with more or less violence, as if an undulation like the ground-swell of the sea were passing beneath. Several shocks in succession, but commonly less destructive than the first, may follow within a few seconds.

5. The earthquake is really of the nature of a wave passing through the substance of the solid earth. Just as the masts of ships at rest in a harbour are seen to rock to and fro, as the swell of the sea rolls under them, so trees and other tall objects have been observed to sway backwards and forwards during an earthquake.

6. On steep banks and cliffs, large masses of clay or rock are sometimes disengaged by an earthquake shock, and are sent rolling down into the water-course or valley below. The streams are thus choked up, temporary lakes are formed, until, by the bursting of the barrier of fallen debris, the water rushes out with great force and carries another kind of devastation down the valley.

7. On many occasions the ground has been seen to be rent open. The fissures thus formed sometimes swallow up trees, houses, or other objects on the surface, and close up again. Sometimes they remain open and are subsequently deepened and widened by running water into ravines and valleys.

8. Again, the ground has been found to have been permanently raised above or depressed below its former level. Thus in the year 1835, during a violent earthquake, which convulsed the coast of Chili, the ground at the island of Santa Maria was suddenly upheaved from

eight to ten feet above high-water mark, so that the gaping sea-shells still adhering to the rock were exposed to the air, where they soon began to putrefy. The valley of the Mississippi was the scene of a succession of earthquakes, from the end of the year 1811 to 1813, at the close of which time, the ground was in many places left with huge yawning fissures, and in certain districts sank down, so as to be converted into lakes, some of which are fifty miles in circumference. Part of this submerged land is called " the Sunk Country." And even now, large trees of walnut, white oak, mulberry, and cypress may be seen ten or twenty feet or more below the water, and a canoe may be paddled among their submerged branches.

9. When the earthquake shock takes its rise under the sea and travels thence towards the land, the greatest amount of destruction is caused. Not only does the solid land rock to and fro, but the waters of the sea are shaken, and sent with terrible force against the margin of the land. From the point of origin of the shock a long, low undulation is propagated over the surface of the sea in all directions. When it reaches shallow water, its front, like that of the bore of a tidal wave in an estuary (Lesson XVII. Art. 30), becomes steep and advances with great rapidity. To those on shore the approach of this wave is seen to be preceded by a retreat of the sea from the beach. Spaces never bare before are now exposed to view as the water retires, but only for a few seconds. The wave, gathering up the retreating water, rushes forward with a front sometimes sixty feet or more in height, not only covering the beach but even sweeping far in upon the land.

10. This sea-wave does not travel across the ocean so fast as the earthquake-shock. Hence it arrives at a coast some time later and completes the destruction. This was the case in the famous earthquake of 1755, by which Lisbon was destroyed. Again, on the 13th of August 1868, when Peru and Ecuador were devastated by a

disastrous earthquake, a great sea-wave inundated and overthrew Arica, the principal port in the south of Peru, with such rapidity and completeness, that in a few minutes every vessel in the harbour was either ashore, wrecked, or floating bottom upwards. A man-of-war was swept inland for a quarter of a mile. Another vessel disappeared, and no vestige of it was ever seen again. Lastly, in the great eruption of Krakatau, in August 1883, a sea-wave, said to have been as much as 100 feet high, swept across the Sunda Strait, and is supposed to have killed between 30,000 and 40,000 people.

11. The true cause of earthquakes is not yet well understood. Probably they take their rise from more causes than one. Thus, they may sometimes be due to the giving way of the roofs of the cavities which no doubt exist in the interior of the earth, especially in volcanic countries; sometimes to the sudden fracture of rocks under great strain; or to the sudden generation or escape of steam. Whatever be the cause of their origin, some sudden blow within the interior of the earth will account for the phenomena of earthquakes. The point directly above that from which the blow is given suffers the severest shock, and the intensity diminishes outwards from this centre. The effect may be compared to that produced on a surface of still water into which a stone is thrown. Where the stone strikes, a series of rings is formed which are propagated outwards in ever-widening circles, but become less marked and farther apart as they recede from the point of origin.

12. If, after an earthquake has passed, the point can be ascertained where the shock was vertical, and if, by further observations upon the direction of the rents in walls and other evidence (Fig. 38), it can be determined at what angle the earth-wave emerged at the surface in one or more places beyond the centre, an indication may be obtained as to the probable depth at which the shock took its rise. Let V in Fig. 39 be the point at which the

shock was felt to have come up vertically, and A a village or town where, from the prevalent direction of the rents in the buildings, the shock appears to have come up

FIG. 38.—House rent by earthquake (Mallet). The arrow shows the direction from which the earthquake-wave must have reached the surface.

obliquely. If we prolong the path indicated by these rents until it meets the vertical, we obtain at F the probable focus of disturbance. Calculations of this kind

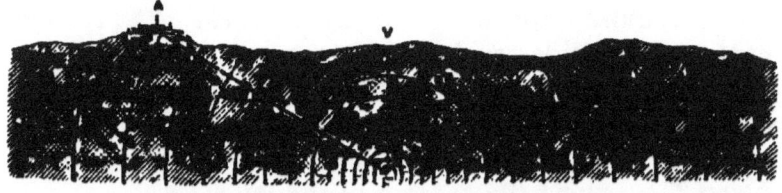

FIG. 39.—Diagram-section to illustrate the propagation of an earthquake-wave and the mode of calculating the depth of its origin.

were made by the late Mr. Mallet, who concluded that, on the whole, earthquakes are not deep-seated, but probably never arise at a greater depth than thirty geographical miles. As implied in the diagram, the successive waves, like the rings on the pool of water (Art. 11), are closer and stronger at the focus, whence they spread in all directions through the solid earth, becoming less and less violent as they recede.

13. Earthquakes, as represented in Plate IX., are most frequent in volcanic districts, though not by any means confined to them. The great earthquake region of the Old World stretches from the Azores along the basin of the Mediterranean into the heart of Asia. In the New World the western border of the Continent suffers most from earthquakes, especially from Guatemala southwards through Ecuador, Peru, and Chili.

14. Most frequently the more striking effects of earthquakes are comparatively local, though pulsations in lakes and on the sea may be noticed at long distances from the centre of disturbance. At the time of the Lisbon earthquake some of the lakes of Scotland were agitated, and the sea-wave that was generated travelled across the Atlantic, and was felt on the American coast. Occasionally the actual vibration of the ground is sensibly felt for a great distance, as in the case of the earthquake of the 13th August 1868, which was felt in Peru for 2000 miles.

15. (ii.) **Slow Upheaval.**—After an earthquake has ceased, the ground is sometimes found to have been permanently raised above or depressed beneath its previous level. But besides such sudden movements others have affected different parts of the earth in so slow and quiet a manner as not to be perceptible at the time. As a rule, they can only be detected by careful observation of their effects along the margin of the land.

16. Old harbours and sea-walls are sometimes found to stand now at some height above even the highest tide. Islands once separated by a water-channel from the land are now joined to it. Caves, evidently hollowed out by the sea, may be seen far above the reach of the

waves. Barnacles and sea-shells are found still adhering to rocks, at heights of several hundred feet above the level where they lived and died. Terraces of sand and gravel, quite like the present beach, and containing sea-shells, occur at different heights above the sea (Figs. 40 and 41). These terraces are old beaches, each of which marks a former level of the sea, before the land rose to

FIG. 40.—View of an old sea-terrace or raised beach, with sea-worn caves on its inner margin.

its present height. Many such terraces skirt the shores of Great Britain. They form a marked feature of the coast in the north of Norway (Fig. 41). On the western margin of South America they occur in great perfection, reaching sometimes a height of 1300 feet above the sea, where sea-shells, still in position, attest the amount of uprise.

17. By evidence of these various kinds it has been ascertained that many long tracts of coast-line are slowly

rising from the sea. Thus the shores of Sweden at Stockholm and northwards appear to be upheaved at a rate varying from six or ten inches to two and a half feet in a hundred years. Farther north, the island of Spitzbergen is fringed with raised beaches up to a height of 147 feet. The coast-line of northern Russia and Siberia for hundreds of miles has been recently elevated out of the sea, as is shown by raised beaches with marine

FIG. 41.—Raised sea-terraces of the Alten Fjord, Norway.

shells. The shores of the Mediterranean afford local illustrations of uprise, while the great sandy desert of the Sahara, containing here and there scattered sea-shells, up to a height of 900 feet, is a case of the recent elevation of a wide tract of sea-bottom.

18 (iii.) **Slow Subsidence.**—While in some regions the slow, imperceptible movement of the ground is an upward one, in others it is downward. This may be shown by different kinds of evidence. Thus, some seaport towns in the south of Sweden contain, under their present streets, traces of older structures which are now

below the sea-level. At different parts of the coast of Scotland and the south-west of England, stumps of trees, with their roots still imbedded in the soil on which they grew, are to be seen actually under the water of the sea.

19. Until recently the most striking evidence of the subsidence of large portions of the earth's surface was believed to be supplied by the **Coral Reefs** of the Pacific

FIG. 42.—Section of an island (L), with a Fringing coral reef (R)

and Indian Oceans. These are reefs or submarine banks of a kind of lime-stone, formed entirely by the growth of coral polypes. Three kinds of reef were recognised by Mr. Darwin. The first, called Fringing Reefs, skirt the margin of the land at a distance of one or two miles, with a shallow-water channel or lagoon inside. The outside of the reef is its highest part, and there the large strong kinds of coral live which delight in the dash and play of the waves; the finer and branched kinds prefer the stiller water of the lagoon. Reefs of this kind skirt the east coast of Africa and occur also in the West Indies and fronting the shores of Brazil. The accompanying

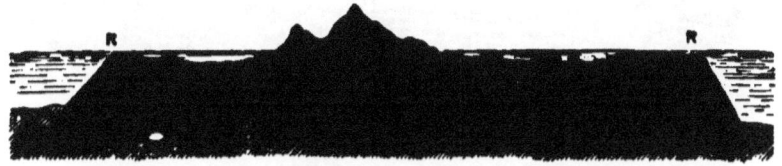

FIG. 43.—Section of an island (L) with a Barrier reef (R).

figure (Fig. 42) shows a section of an island where the land (L) is skirted by a fringing reef (R).

20. The second kind of reef is called a Barrier Reef (Fig. 43). It may skirt a long tract of coast, as in the

north-east of Australia, where the great reef fronts the land for 1000 miles; or it may encircle an island, as at Tahiti. Its slope to the sea is in the upper part nearly precipitous, but beneath this steep face the bottom slopes away gently into the surrounding sea-floor. Inside, the reef is separated from the land by a deep lagoon channel.

FIG. 44.—Section of an Atoll or coral island (R R) built over submerged land (L).

There may sometimes be more than one island inside a barrier reef.

21. The third form of reef has received the name of Atoll or coral island (Figs. 44 and 45). It is a ring of coral rising out of a deep ocean, and having a

FIG. 45.—View of an Atoll or coral island.

breadth of about a quarter of a mile. The outer face towards the sea is precipitous in the upper part, like a barrier reef. Inside lies a lagoon of comparatively shallow water, full of the more delicate branching kinds of coral. The waves break off fragments from the outer edge of the reef, and pile these up above the ordinary limits to which the sea reaches. By degrees a narrow ring of land is formed on the reef. Seeds are carried to it and take root, and at last a human population arrives and finds shelter and the means of subsistence.

22. It has been ascertained that the reef-building corals cannot live in deep water, that indeed they do not go down farther than about twenty fathoms. Consequently they cannot have grown up from the deep bottom on which the atolls are planted. As they could not have begun to build more than about twenty fathoms from the surface, Mr. Darwin inferred that the bottom has been gradually sinking, while the little coral-builders have kept pace with the subsidence, and have maintained their reef at the sea-level. The three sections here given (Figs. 42, 43, and 44) illustrate the stages in this supposed subsidence. First comes the fringing reef, with its shallow lagoon and not very deep water outside. Then this shore reef passes into the barrier reef, with its deeper lagoon channel and much deeper sea outside. The land inside becomes less in height and extent as it settles down beneath the sea-level. At last it sinks out of sight altogether, and the barrier reef remains as an atoll.

23. This theory of Mr. Darwin's has been so long accepted, and presents so impressive a picture of vast subsidence in the ocean-basins, that it still holds its ground. More recent observations, however, by Messrs. Semper, Murray, Rein, and Agassiz, have brought to light many facts that were not known to Mr. Darwin, and which show that the growth of coral-reefs and islands can be satisfactorily explained without the supposition of any depression of the sea-bottom. In particular, it has been ascertained that in the surface water and on the bottom of tropical seas animal life is so prodigiously abundant, that the remains of the calcareous organisms accumulate on the bottom as a growing deposit of limestone. By the continued growth of such limestone the tops of submarine

peaks and ridges may eventually be brought up to the limit of depth within which reef-building corals can thrive. Having once established themselves on the platform prepared for them, these creatures grow upward until they reach the level of low-water. But, though unable to grow higher, they continue to thrive vigorously on the outer face of the reef in the full play of the surf, which brings them their supplies of food. The breakers tear off huge blocks of solid coral and strew these down the slope below, piling them here and there even into a precipitous wall. Meanwhile the living coral continues to build outward on the top of the blocks which, by the chemical action of the sea-water and the settling of coral-sand into their interstices, are cemented together into a compact mass. In this way, the steep sea-ward front of barrier reefs and atolls may be explained. In the inner parts of a reef the corals, being removed from contact with the open sea, from which they get their food, dwindle and die. Hence the coral rock does not grow in the lagoons. On the contrary it appears to be gradually eaten away by the solvent action of the sea-water, whereby the lagoons are probably widened and deepened.

LESSON XXIV.—*The Waters of the Land—Part I.*

Springs and Underground Rivers.

1. From every water-surface on the globe invisible vapour is ascending into the air, where it is condensed into clouds, and whence it is returned to the surface of the earth again in rain, dew, snow, hail, or sleet. Discharged upon the land, the water partly sinks underground to rise again in the form of springs, partly flows off in rivers into the sea, whence, once more evaporated, it enters the atmosphere to begin again the same cycle of change (Lesson X.)

2. There is thus a continual circulation of water between the atmosphere, the land, and the sea. The more this circulation is considered the more importance will be assigned to it in the general plan of the earth. The clouds form and melt and form again. Day by day, or season by season, the rain-showers reappear, to moisten the parched soil. The brook and waterfall, ever rushing downwards, are yet continually fed with renewed supplies from above. The river still bears its broad breast of waters to the sea, as it has done ever since the earliest tribes of men settled upon its banks. The sea, though receiving the surplus drainage of all the continents, is not thereby raised in level, but yields to the air those abundant vapours which, borne back to the land, and condensed into running water, once more renew their downward journey to the sea. This continual coming and going of water may be looked upon as the pulsation of the very life-blood of our globe as a habitable planet.

3. Let it be remembered, too, that water enters largely into the composition of the bodies both of plants and animals. If its circulation were arrested, our earth would cease to be the green, populous planet which it is at present. Were evaporation and condensation to cease, clouds, springs, and rivers would disappear. Scorched by the fierce heat of the sun during the day, and frozen by the intense cold of radiation at night, the land would become lifeless and silent.

4. The moisture of the air returns to the land either in the liquid form, as water, or in the solid form, as ice (Lesson X.) In the former state it chiefly appears as Rain; in the latter as Snow. The all-important part which, under each of these conditions, water takes in the daily operations of nature, deserves the most attentive study. In this and the following Lessons we shall look more in detail at the circulation of water over the land, taking first Rain and its consequences, and then Snow.

5. When rain reaches the surface of the land, part of it sinks into the soil, and the rest flows off into brooks and rivers, by which it is carried back to the sea. These two portions of the rainfall have each an independent history, and may be conveniently considered separately. In the present Lesson, therefore, we shall deal with the underground course, and in the following Lesson with the course above ground.

6. It might naturally be supposed that the portion of the rain which sinks into the earth is permanently withdrawn from the circulation of water on the earth's surface. But a little reflection will convince us that if this were really the case, the amount of water flowing over the land would be diminished. Rivers and lakes would shrink in size, or dry up altogether. Yet as this does not happen, there must obviously be some way in which the water that sinks into the ground is restored to the surface again. This restoration takes place by means of **springs**, which are the outflow of the subterranean water from openings in the ground.

7. The intimate connection between ordinary springs and rainfall is familiar to every one. We know that in a season of drought many springs and wells give a limited supply of water, or fail altogether, while, as wet weather sets in, they fill again. They obviously derive their supplies from rain-water, which has percolated through the rocks beneath the surface. Such springs as have a deep-seated origin are less affected by rainy or dry seasons, because they gather their stores from a wider area of subterranean drainage, where the effects of a scarcity of rain take longer to make themselves felt than is the case near the surface.

8. All rocks, even the hardest, are porous, and therefore pervious to water. The channels of brooks and rivers, the beds of lakes, and the floor of the sea, are all more or less cracked, so as to present openings for the descent of water. Rain-water, therefore, not only soaks

through the soil, but, sinking lower still, finds its way through the pores and joints of rocks underneath. Water is likewise supplied by lakes, rivers, and the sea, either oozing through the pores of rocks or entering open cracks, and carrying with it sand and other impurities.

9. In the sinking of deep wells in some districts of France, leaves and other parts of plants have come up with the first gush of water from a depth of nearly 400 feet. These organic remains were comparatively fresh, and were supposed to have travelled in underground channels from hills 150 miles distant, and to have been three or four months on their subterranean journey. The same phenomenon has been observed in other places; sometimes even live fish have been brought up in borings from depths of 170 feet.

10. As the result of this constant percolation and descent of water from the surface, the rocks, for some way down, are in many places charged with moisture. Proofs of the almost constant presence of water are furnished in quarries, pits, and mines, in short, in nearly every place where any considerable cutting is made through rock. It is this underground water which forms one of the greatest obstacles in quarrying and mining operations. Before the introduction of steam machinery many coal-pits, after having been worked to a certain depth, had to be abandoned, from the impossibility of getting rid of the water. They were then said, in the expressive language of the miners, to be "drowned." The powerful pumping engines, now to be seen everywhere in the coal-fields, point to the abundance of water below ground, and to the labour and cost which are necessary for removing it.

11. Another and familiar illustration of the way in which water everywhere pervades the soil and rocks is to be seen in the sinking of wells. These are artificial cavities, dug out to serve as receptacles wherein the water that is soaking through the rocks may be collected.

Wells may be successfully made even in places where it could hardly be supposed that water would be found. Thus, on the borders of the African deserts, where little or no rain falls, and where, therefore, there can be but a scant supply of water from the surface, serviceable wells are dug. The French colonists of Algeria sink what are known as "Artesian wells" (Art. 18) on the northern margin of the great desert of Sahara, and the sandy tracts between Cairo and Suez yield water even so near the surface as at a depth of fifty feet. The existence of those fertile green patches called Oases, in the midst of the barren deserts of Africa and Arabia, is due to the rise of springs. Again, in the valley of the Mahanadi and other Indian rivers, where, in the dry season, little or no rain falls, a hole dug out to the depth of thirty or forty feet is sure even then to be partly filled with water.

Hence we may conclude that the springs of a district do not always or necessarily obtain their water from the rainfall of the immediate neighbourhood. If that were the case there could hardly be perennial springs and wells in the African deserts, where rain is exceedingly rare.

12. To what depth the water will descend must depend greatly upon the nature and condition of the rocks at each locality. Very deep mines are often without water. When the Alps were pierced in making the railway tunnel between France and Italy, the rocks at a depth of more than 5000 feet below the summit of Mont Cenis were quite dry. We need not suppose, therefore, that the water generally sinks to a very great depth. But here and there it no doubt does find its way down even into the intensely hot internal regions of the earth, whence it reissues in those vast clouds of steam that play so important a part in the arrangements of active volcanoes (Lesson XXII.). It is probable that in spite of the plentiful discharge of steam at these volcanic openings, a part of the water which descends so far may be

lost by absorption into the molten rock, or by being decomposed, and forced to enter into chemical combination. If this be the case, then the earth must be losing its superficial water; slowly and insensibly, indeed, but yet if continuously, with the probable result of ultimately reducing our planet to the dry and sterile condition of the moon.

13. Though the water which falls upon the land is distributed over the surface as rain, when it reappears at the surface it does not ooze everywhere from the soil. Sinking underground, it finds its way along cracks and hollows of the rocks below, until it comes out again to the surface at certain points. Just as in the subaërial course of the fallen rain, the water at once runs off into brooks and larger streams until it finally enters the sea, so in a somewhat similar way, the underground drainage, collected from many branching channels, is brought out to the surface in springs.

14. A difficulty may be felt in understanding how the water, having once sunk down, can ever be driven up again. But we must remember that the springs, which form its points of escape at the surface, lie at a lower level than the ground from which the original supplies of rain have been drawn. A little reflection on this subject will convince us that the underground circulation must be effected in one or other of two ways; either by simple gravitation, as in what may be called Surface Springs, or by hydrostatic pressure, as in Deep-seated Springs.

15. (I.) In the case of **Surface Springs** the water, which has been steadily flowing downward as well as onward in its underground course, comes to a point where, owing to some depression of the surface, it finds itself again in the open air. The subjoined woodcut will explain how such springs arise. A porous bed of rock (b), or one traversed with cracks or joints, lies nearest the surface, and allows the rain water to soak through it,

down to a stiff impervious layer (*a*), by which the descent of the water is arrested. Unable, therefore, to sink farther downward, the water flows along the surface of this lower bed. If a valley should happen to cut through these rocks, there will be a spring or line of springs (*s*) on the junction of the two rocks along the side of the valley. In the same way, the rain which falls upon a mountain may sink underground, and gush out in springs at the foot of the mountain. In springs of this kind, the water merely descends in the ordinary way by gravitation, and issues where the surface happens to sink below the level of the subterranean water-channel. In heavy rain a good deal of water soaks through the soil into the nearest brooks without ever forming actual springs.

FIG. 46.—Section across a valley to show how the simplest kinds of springs arise.

Wherever water accumulates among underground rocks, whether in the pores or in the open crevices, so as to convert the rocks into subterranean reservoirs, it will rise in them up to the lowest levels at which it can find outlets to the surface, where it will appear in these surface- or land-springs. This "water-level" must be reached before a well can catch a supply of water.

16. (II.) In **Deep-seated Springs**, on the other hand, the water has in its journey sunk to a lower level than its point of escape, and has risen again by hydrostatic pressure, as in a syphon. Evidently the arms of a syphon may be as long as we choose to make them, yet while the one is longer than the other, and is supplied with water at the top, the water will continue to flow out from the top of the shorter arm. In like manner the subterranean channel of a deep-seated spring may descend

for hundreds of feet, yet the water which fills it will not cease to flow and to rise again to the surface. Having reached its greatest depth, often far below the level of the sea, the water accumulates there, saturating such porous rocks as lie in its way, until the pressure of the column of water behind it forces it up any fissure which may allow of its escape to the surface, and there it bubbles up as a spring. The rain which falls on the high grounds, and is absorbed by the rocks and soils, flows down more or less permeable rocks which may be arranged in different beds or *strata* (*a* in Fig. 47). Taking advantage of the cracks that may opportunely

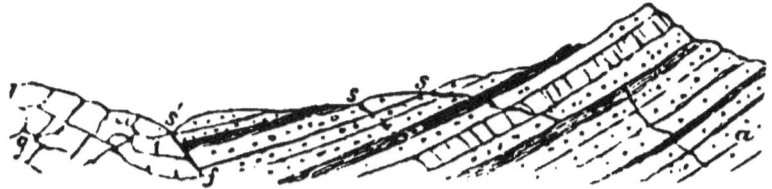

FIG. 47.—Section to show how deep-seated springs arise.

present themselves, as much of the water as these openings will admit rises through them to form springs, as at *s, s*. At *f* the strata are broken across by a great fracture or dislocation, which brings them against a hard massive rock (*g*). Here a considerable body of water may escape to the top at *s'*. Hence the direction of a great dislocation of the rocks is often traceable at the surface by a line of springs.

17. In nature, the course which water takes underground is in reality, for the most part, much more complicated than is shown in the diagram (Fig. 47). All rocks are abundantly traversed by divisional planes called "joints;" they are likewise full of cracks, and they present many changes in texture. So that subterranean water finds a most intricate network of passages through which it must flow. It may possibly come many times

near to the surface, and then sink down again through other fissures, performing in this way a zigzag up-and-down journey, before it finally issues in springs. A rough representation of this kind of circulation is given in Fig. 48.

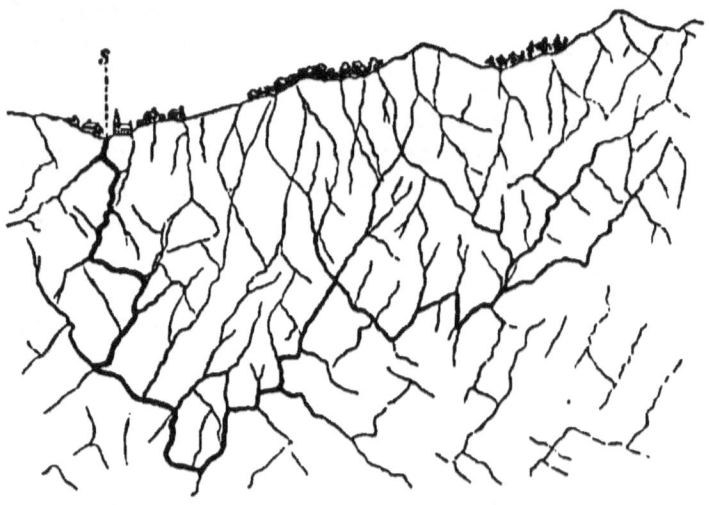

FIG. 48.—Section to show the intricate underground drainage which issues in a deep-seated spring. The numerous joints and cracks in the rocks lead the water at last into a main channel, by which it reascends to the surface as a spring at *s*.

18. On further reflection we perceive that there must necessarily be many underground rocks which are permanently saturated with water. These, if they can be reached, will furnish an abundant and constant supply of water. Advantage is taken of this kind of knowledge to sink deep wells, called "Artesian," from the old French province of Artois, where they have long been in use. The principle on which these wells are made is as follows:—When an impervious rock covers a porous one over a considerable district of country, the water that soaks into the lower bed will tend to accumulate there as in a great reservoir. If now a hole be drilled through the upper retentive bed, the water will rise in it at once,

as it would do if a natural fissure existed there, and may ascend even above the surface of the ground. The water occasionally issues from these bore-holes with such force as to form a jet, rising to a height of thirty feet or more above the soil. In the north of France the force of the jet has been made use of to drive mills. Many wells on this principle have been sunk in London and its neighbourhood. The section (Fig. 49) across the district in which London lies, shows roughly how the water which falls on the high grounds to the north and south, filters through the sands and gravels (*b*) at the top of the chalk (*a*), and is imprisoned by the retentive London clay (*c*).[1] When wells are sunk down to this water-bearing zone, water ascends in abundance. So many

FIG. 49.—Section to the position of the water-bearing rocks below the clay at London.

wells, however, have now been made, that the level of the water in them has gradually sunk year by year, the consumption being thus somewhat more rapid than the supply to the underground reservoirs.

19. In trying to picture to ourselves the amount of water which is always circulating under ground, we ought not to measure it merely by what is seen coming out at the surface in well-marked springs. For, in the first place, we must bear in mind that the abundance of springs is really much greater than might at first sight appear. Much of the water which rises from under ground does not bubble up in the form of well-marked springs. When it reaches the surface, it soaks through the soil or trickles over it in tiny runnels. On unculti-

---

[1] In the diagram the curve of the beds has necessarily been greatly exaggerated, there being in reality hardly a basin at all.

vated land such places are marked by greener patches of vegetation, or by tufts of rushes and a swampy soil. Sometimes, too, we may see them exposed, even on ploughed fields, during dry weather in spring-time. From want of rain the bare soil becomes dry and light in colour. Here and there, however, it is diversified with dark brown patches, which point to places where the water is oozing out from below, and soaking through the soil, in spite of the drain-pipes of the farmer. In these and similar cases, the water, after coming up to the surface, sinks down again into the rocks, and commences another underground journey.

20. In the second place, we must remember that the natural flow of water from a higher to a lower level must carry much of the underground drainage to the rocks under the sea, so that, if the water rises to the surface, it will do so on the sea-floor, instead of under the open air. Springs along the beach within tide-marks are sufficiently common, and many cases have been observed in the Mediterranean where powerful springs or even underground rivers rise to the surface of the sea at some distance from the shore, so that fresh water may actually be obtained at these places for the supply of ships.

21. In some countries, where there are no rivers, and little or no rain, springs and artificial wells are the only sources from which a supply of water can be obtained by the inhabitants. But even in regions where both rain and rivers abound, the usefulness of springs is hardly less apparent. Think for a moment what would happen if all the rain were immediately to run off the surface of the land, without allowing any portion to sink under ground. Except where fed from melting snow, the brooks and rivers would flow only after showers. Their channels would dry up in the absence of rain. It is the underground circulation which, by means of springs, replenishes the water-courses of the land, even in drought, and thus preserves the surface of the land fresh and green.

22. Rain is water nearly in a state of chemical purity; but in its descent it takes up a little air and some of the impurities which may be floating in the air (Lesson VI.). These admixtures, however, form only a minute proportion in rain-water, especially at a distance from the smoke and vapours of towns and manufactories. But the water of any spring, even the clearest and most sparkling, though it sank into the ground as pure or almost pure water, reappears again with an impregnation of various substances. These are often so abundant as to be made readily visible after the water has been boiled and evaporated, when they remain behind as a film on the bottom of the vessel. They are chemically dissolved in the water, and do not interfere with its clearness and brilliancy, nor in most cases do they impart to it any taste. They may be found in the water of every spring, but their quantity varies greatly. Sometimes they occur in such minute proportions as fifty parts in every million parts of water; in what are called mineral waters they may rise to as much as 32,700 parts in the million.

23. Three questions naturally arise in reference to this remarkable impregnation of all spring-water: firstly, what are the substances present in the water? Secondly, how does the water obtain them? and, thirdly, what is the result of their constant removal by the water?

24. (i.) Substances contained in spring-water. If a glass of bright sparkling spring-water be allowed to stand for a time, minute bubbles may be observed adhering to the inner surface of the glass. These are air, or gas, which has been dissolved in the water. They rise by degrees and escape, and when they have done so, the water ceases to have the same fresh brisk taste which it had when first drawn from the spring. In like manner, if the water is boiled it acquires an unpleasant insipid taste, not because anything has been put into it, but because the air or gas which it contained has been driven out. Hence it appears that at least one cause of the pleasantness of spring-water to the taste is the presence of gaseous materials.

25. Chemists have carefully examined these materials, and have found that they consist mainly of oxygen and nitrogen, that is, the ordinary gases of the air. Pure water when saturated with air will contain by volume 17·95 parts of air in every 1000 parts of water. The proportions of the two atmospheric gases, oxygen and nitrogen, in this dissolved air are considerably different from those in the atmosphere, being in every hundred parts 65·09 of nitrogen and 34·91 of oxygen. The latter gas being more soluble exists in considerably larger proportions in water than in the air (Lesson X. Art. 35). Spring water also contains carbonic acid, and sometimes sulphuretted hydrogen, marsh gas, or other gases.

26. The fine white film of mineral substance left behind, when a quantity of spring-water has been boiled down and driven off into vapour, is found to vary greatly in composition in different springs. It commonly consists of carbonates or sulphates of lime, soda, magnesia, and often includes chloride of sodium or common salt. In limestone districts, owing to the large proportion of lime or magnesia in the water, much soap is needed in washing, because the mineral constituents unite with the fatty acid of the soap to form a curdy insoluble compound. Such water is said to be *hard*. When the hardness is due to the presence of carbonates, it disappears when the water is boiled. A hard crust is deposited on the inside of the kettles or boilers in which the water is boiled. If the hardness arises from the presence of sulphates it cannot be removed by boiling. Where the percentage of mineral matter is small, the water is called *soft*. The softest water that can be had for domestic purposes is of course rain, for it contains little or no mineral admixture.

27. While all spring-water holds more or less mineral matter in solution, some springs contain so much that it affects the taste or becomes visible as the water flows over the ground. These are known as **Mineral Springs**. The chief mineral substance in solution is carbonate of lime. When the water begins to evaporate, as it trickles away from the spring, the lime is sometimes deposited as a white crust upon stones or other objects lying in its way. This is particularly apt to occur on certain kinds of moss which prefer water containing much lime. A white limey crust gathers round every little fibre and completely encloses it, so that as the plant dies the white crust remains and retains the outward form of the original moss. These *calcareous* springs, as they are called, occur abundantly in limestone countries. In other springs, carbonate or sulphate of iron is the principal substance contained in the water, which, as it flows along, deposits iron oxide as a yellow or brown scum on the sides of its channel. Springs of this nature are known as *ferruginous* or *chalybeate*. Common salt is the characteristic of some springs, and when they contain a large proportion of it they receive the name of *brine*-springs. Sometimes they are nearly or quite saturated with salt. Many springs containing a considerable admixture of mineral ingredients are useful in certain diseases, either in the form of draughts or of baths. They are termed *Medicinal* springs. For example, the waters of Bath contain carbonate of lime, sulphate of soda, sulphate of lime, chloride of sodium, silica and carbonic acid; those of Harrogate, carbonates of magnesia and lime, sulphate of magnesia, chlorides of sodium, magnesium, and lime, nitrogen, carbonic acid and sulphuretted hydrogen. The waters of Vichy are alkaline and acidulous from the quantity of soda, potash and carbonic acid which they contain. Those of Carlsbad abound in sulphates, those of Wiesbaden in chlorides.

28. In mountainous and snow-covered ground, spring-water may be met with scarcely above the freezing-point. From that extreme, every degree of temperature may be

noted in different springs, up to such boiling fountains as the Geysers. Evidently, the temperature of the spring must depend upon that of the rocks from which the water rises. If the water comes from melting snow and ice, it will necessarily be cold. If it sinks deep within the earth, so as to come within the influence of the internal heat, it will be warm. So that there seems good reason to regard the temperature as affording some indication of the probable depth from which the water has risen.

29. Hot water has a greater power of dissolving most substances than cold water. Consequently, warm or thermal springs are often largely impregnated with mineral matter. Many medicinal wells are warm. Those of Vichy, for example, have a temperature of 111° Fahr., those of Carlsbad, 165° Fahr., and those of Chaudes Aigues, 178° Fahr. One of the substances particularly soluble in hot water is silica, which has been already alluded to as occurring in the water of boiling springs, and as being deposited in the form of a hard crust round the orifices from which the water escapes, as in the Geysers of Iceland, the Yellowstone Park and New Zealand (Lesson XXI. Art. 24).

30. The proportion of mineral matter held in solution differs widely in different springs. In common spring-water it may range from 50 to 400 or 500 parts in every million parts of water. But in districts where the water is "hard" the proportion may rise to 2000 parts in every million. In mineral springs, the proportion is of course very much greater. Thus, in the Vichy waters, the solid contents are $\frac{5160}{1000000}$, those of Seidlitz contain $\frac{10400}{1000000}$; those of Püllna, in Bohemia, $\frac{32771}{1000000}$. The proportion in the Atlantic Ocean water, is about $\frac{35000}{1000000}$, while in the Dead Sea, it is $\frac{240483}{1000000}$.

31. When the discharge of water, as well as the proportion of dissolved mineral matter, is large, the quantity of material dissolved out of the rocks below ground and carried up by springs, must be enormous. Thus, one of the hot springs of Leuk, in Switzerland, having a temperature of 144° Fahr., is estimated to bring to the surface every year nearly 9,000,000 of pounds of gypsum. If this mass of mineral could be collected it would form a square column, upwards of 650 feet high, and twenty-seven feet on each side. The brine spring of Neusalzwerk, near Minden, has been found to yield in one year enough of brine to form a cube of solid salt, measuring seventy-two feet. The wells of Bath, in like manner, are computed to yield annually enough of mineral matter to form a square pillar ten feet on the side, and eighty feet high.

32. (ii.) **Origin of the Substances dissolved in Spring-water.** Rain in falling dissolves a little air, but takes considerably more oxygen and carbonic acid gas than the normal proportions of these gases in the atmosphere (Lesson VI. Art. 35). The proportion of carbonic acid gas indeed is between thirty and forty times greater. In sinking through the soil rain obtains, from decomposing plant and animal remains, more carbonic acid and various organic acids. In old volcanic districts, as Auvergne in Central France, and the Eifel in Rhenish Prussia, carbonic acid is given off in great quantities from subterranean sources, and is here and there brought up by springs, which like the Bad Tönnistein in the valley of Brohl, are so full of it that it escapes in copious bubbles when the water comes to the surface.

33. The oxygen and carbonic acid carried by percolating rain downward into subterranean rocks are of great importance in aiding the water to decompose rocks. The presence of carbonic acid enables it to dissolve large quantities of various mineral substances. How this solution takes place is admirably illustrated above ground by what happens at the arches of many bridges. On the under side of an arch, or the vaulted roof of a cellar, each line of mortar between the courses of masonry is often marked by a sort of fringe of

slender pendent white stalks or pencils. At the point of each of these stalks, a clear drop of water may be noticed, which in time falls to the ground, and is slowly replaced by another drop. Should the ground beneath the arch be undisturbed by passing footsteps, where the drops fall upon it, a white deposit, like that of the roof above, grows up in little mounds, each of which is kept wet by the constant drip of the water. These white incrustations may be seen to increase in size from year to year. It is evident that they have come out of the masonry and that trickling water has had something to do with their formation.

34. Rain-water containing, as it does, carbonic acid, has the power of dissolving lime, and holding it in solution in the form of what is known as carbonate of lime. The mortar which binds the masonry together consists mainly of lime. Being usually more porous than the stone or brick, which it cements, it allows some rain to sink through the seams and joints of the masonry. The water in its passage from the upper to the under surface of the archway, takes out a little lime, and carries it off in solution. As each drop appears upon the roof and hangs for a time before falling, it is diminished by evaporation, and losing part of its carbonic acid, cannot hold so much lime as at first. It is therefore compelled to deposit the surplus as a white film upon the roof. The drop then falls, and is succeeded by the next, which goes through the same stages, and thus the original ring of lime, left round the edges of the first drop, grows into a long slender hollow tube or stalk, like an icicle of stone. If it is undisturbed, it may lengthen until it actually reaches the floor, and its sides may be added to by further trickling water until it becomes a stout rod, or even a pillar seeming to support the roof. These hanging stalks or columns of lime are called *stalactites*.

35. But the drops do not leave all their lime behind them on the roof. They carry some of it with them to

the ground, where, on further evaporation, they deposit it as a white solid crust called *stalagmite*. This process of removal and re-deposit, seen so well in a small way in arches of masonry, occurs sometimes on a magnificent scale in vast limestone caverns (Arts. 39-41).

36. If, now, in sinking through a few feet of masonry, rain-water can work such great changes, what may we not expect to take place under ground, where the water has to traverse vast masses of rock, and where its solvent power may be increased by the earth's internal heat!

37. (iii.) **Results of the removal of materials from below in spring-water.** In tracing the results of the universal percolation of water through underground rocks, and of the removal and transport to the surface of so much of their solid substance, we may consider first the effects on the surface, and secondly the effects below ground.

38. With regard to the influence exerted on the earth's surface by the solution of subterranean rocks, we may reflect that springs supply rivers and lakes, and thus, indirectly, the sea, with lime, iron, soda, and other soluble ingredients. In fresh water and in the sea, there are large tribes of animals which obtain the materials of their shells or skeletons from the lime or other minerals present in the water. If these mineral substances could be removed for a time, the result would be to destroy a large part of the denizens of our rivers, lakes, and seas. Man himself would suffer, not only by losing some of the most valuable of his supplies of food—molluscs (oysters, etc.), crustaceans (lobsters, crabs, etc.), and fishes, but from the mere absence in his drinking water of mineral substances, useful in keeping his body healthy. The failure of medicinal springs, too, would deprive him of an important element in the curative treatment of many diseases.

39. It is in its underground operations, however, and the effect of these upon the surface, that the effects of the

removal of mineral matter by springs are most striking. Since every spring is continually dissolving and carrying up to the surface some of the solid substance of the earth's interior, and since the amount so carried by many springs even in a single year would, if it were collected in the solid form, make considerable hills, the result must be that cavities arise among the subterranean rocks. That this is really the case is best seen in limestone countries. Limestone is liable to be dissolved and removed by percolating rain-water, in the same way as the mortar of an archway (Art. 33). It is full of chinks and joints, by which water finds its way downward. Each of these passages is gradually widened by the solution and removal of the rock at its sides. Hence in districts where limestone forms the uppermost rock, the ground is sometimes full of holes, which have been eaten out of the solid rock by trickling rain-water, and which lead down into numerous branching tunnels and chambers underground excavated in the same way.

40. One of the most remarkable examples of this kind of scenery is that of the Karst in Carniola on the flanks of the Julian Alps. It is a table-land of limestone, so full of holes as to resemble a sponge. All the rain which falls upon it is at once swallowed up and disappears in underground channels, where, as it rushes among the rocks, it can be heard even from the surface. Some of the holes which open to daylight lead downward for several hundred feet. Some turn aside and pass into tunnels in which the collected waters move along as large and rapid subterranean rivers, either gushing out like the Timao at the outer edge of the table-land, or actually passing for some distance beyond the shore, and finding an outlet below the sea. Here and there the labyrinths of honeycombed rocks expand into a vast chamber, with stalactites of snowy crystalline lime hanging from the roof, or connecting it by massive pillars and partitions with the floor. Such is the famous grotto of Adels-

berg near Trieste—a series of caverns and passages with a river rushing across them.

41. Still more extensive is the Mammoth Cave of Kentucky—a cavern about 10 miles long, but with many ramifying passages which are said to have a united length of more than 200 miles. In the Island of Antiparos, a famous grotto lies 600 feet below ground, forming a spacious hall 300 feet wide and 240 feet high.

42. Partly from the solvent action of descending water upon the sides of chinks of the limestone, and partly from the falling in of the roofs of underground passages, the surface of the ground in some limestone countries is so full of holes, and at the same time so bare of soil, as to become a kind of barren and dry desert. All its rain, its springs and its rivers are withdrawn from the surface. In the course of time the ground has here and there subsided to such an extent as to form hollows in which the water collects into lakes. These, however, have no outlet at the surface. The water issues from openings in the rocks to fill them, and flows away by other openings of the same kind. The Zirknitz See in Carniola is a good instance of a lake of this kind. It is about five miles long and from one to two miles broad, but usually not more than from six to ten feet deep. Its bottom is said to be perforated with about 400 funnels or pipes through which the water ascends. In wet weather it rises to three times its ordinary height. But even at high water, it is so surrounded with high ground, that it cannot find any outlet at the surface, and has to discharge its surplus waters down some of the innumerable caverns in the limestone (Lesson XXVI. Art. 13).

43. In some parts of the world, therefore, and more especially in limestone regions, the surface of the ground is greatly changed by the chemical action of subterranean water. But there is yet another way in which the circulation of water below ground affects the form of the surface. Where rain sinks through a porous sloping bed of soil or rock, it sometimes forms a loose watery layer underneath, which, by destroying the support of the overlying mass, allows the latter to slip down the

Fig. 50.—Section across the cliff and landslip of Antrim.

slope and tumble into fragments below. This is called a **landslip**. Changes of this kind can, of course, only occur on the sides of mountains, cliffs, ravines, or steep

Fig. 51.—View of part of the cliffs and landslip of Antrim.

slopes, where movement by gravity from a higher to a lower level is possible. They are common along the sea-coast, many parts of the shore-line of the British islands being fringed with old landslips. When the slipped mass is large in extent and becomes covered with vegetation, it forms a strip of broken and picturesque ground in front of the higher cliff behind. Such is the undercliff of the Isle of Wight, and the long line of rough crags and grassy mounds flanking the steep cliffs of Antrim. In the latter case (Figs. 50, 51) a great tableland of ancient hard lava beds (b) rises from the coast in a line of noble cliffs, resting upon layers of much softer and more porous rocks (s). Owing to the loosening of the support of the upper part of the cliff by the trickling of water between the beds in the lower half, huge slices of the heavy solid lava-rocks have been launched down to the low ground. Many of these fallen fragments are themselves large enough to be called hills.

44. In hilly countries subject to heavy rains, landslips are of frequent occurrence. The earth and upper layers of rock, saturated with water, are loosened, and slide down the slopes, carrying trees and fields to the valleys below, and piling up vast heaps of ruin there. In Sikkim, and other districts to the south of the Himalayan chain, the surface of the ground is being altered from this cause after every heavy fall of rain, vast spaces of mountain-slope, many acres in extent, being detached so as to sweep down, with their covering of forest, into the lower ground. Sometimes these fallen masses of earth and rock are thrown across a valley, so as to bar back the river and form a lake. But as the barrier consists only of loose rubbish, it is apt to give way to the pressure of the accumulating waters, which then pour down the valley with great force, sweeping everything before them, and desolating the district for many miles along their course.

45. When landslips take place in well-peopled and cultivated valleys, they sometimes cause great destruction of life and property. Thus, in the valley of Goldau, Switzerland, in the year 1806, after a continuance of wet weather, a bed of rock, 100 feet thick, resting on saturated sandy layers, slipped down. The whole side

of the mountain of the Rossberg seemed to be in motion. In a few minutes the descending mass had, with a terrible noise, rushed across the valley, burying five villages and about 500 people under a mass of ruined rocks 100 to 250 feet high.

LESSON XXV.—*The Waters of the Land—Part II.*

Running water.—Brooks and Rivers.

1. Having in the last lesson followed the course of that portion of the rainfall which disappears into the ground, we have now to trace what becomes of the rest. Since the surface of the land is higher than the sea, and slopes downward to the sea-level, the water which falls upon it from the sky cannot remain there, but must, in obedience to gravity, seek the lowest level. This it can only do by moving downward, till it flows into the sea. If the land sloped evenly from a central ridge, like the roof of a house, the rain might run off in sheets of water. But instead of such uniformity it everywhere presents the most irregular surface. Even on what we might suppose to be a perfectly level and smooth piece of ground, there are innumerable little heights and hollows, which become at once apparent during a shower of rain, for then the hollows are marked by tiny runnels of water which course along them to a lower level.

2. **Water-courses and River-basins.**—Owing, therefore, to the unevenness of the surface of the land, the surplus rain runs off into the hollows, down which it flows until it can descend no farther. These hollows or channels, which receive and conduct the drainage of the land, are termed **water-courses**. They vary in size, from the rut that holds the merest rill or gutter, up to the bed of a broad river which carries the drainage of half a continent to the sea.

3. Take the map of any large country, or of a continent, and notice in the arrangement of the water-courses upon it, that they are grouped somewhat like the branches and stem of a tree. On the low grounds, towards the sea, a river, in one single trunk, bends to and fro across the land. But farther inland it divides into separate limbs, these again into smaller branches, and so on as far as the upper limit of the region drained by the river. This is the common plan on which the drainage of the land is arranged,—innumerable little runnels in the heart of a country, coalescing more and more as they descend, until they all finally unite into a few main streams.

4. Let us in imagination trace the course of a large river, from its beginning in the midst of a continent, to its end in the sea. Among the far mountains, where the sources of the river must be sought, the higher summits are probably covered, or at least streaked, with snow, while long tongues of snow and ice may be seen creeping down the upper parts of the valleys. Perhaps the river issues from the melting end of a "glacier" (Lesson XXVIII.). If so, it springs up at once as a tumultuous torrent of muddy water, and rushes down the valley, receiving from either side innumerable minor torrents and runnels, which descend the rugged slopes, either from the melting edge of the snow, or from abundant clear bubbling springs. Or perhaps the river takes its origin in some single spring, not larger, it may be, than many others in its neighbourhood, but which has been fixed upon from early times by the human population of the district as the true fountain of the river. Such a spring, either welling quietly from the ground or gushing out copiously, supplies the first little stream which dashes down its rocky channel, receiving from each side, as it descends, many tributary torrents, until after leaping from rock to rock in foaming cascades, and working its way through deep gullies, it reaches a more

level part of the valley. In this first or torrent part of its course, the infant river is only one of many such streams by which the sides of the higher mountains are channelled.

5. But when it gains the valley it enters on a second and distinct portion of its journey. Its flow is less rapid, its channel less steep and uneven. It winds to and fro, in many bends and loops, across the flat parts of the valley, and rushes through the narrow gorges that occur

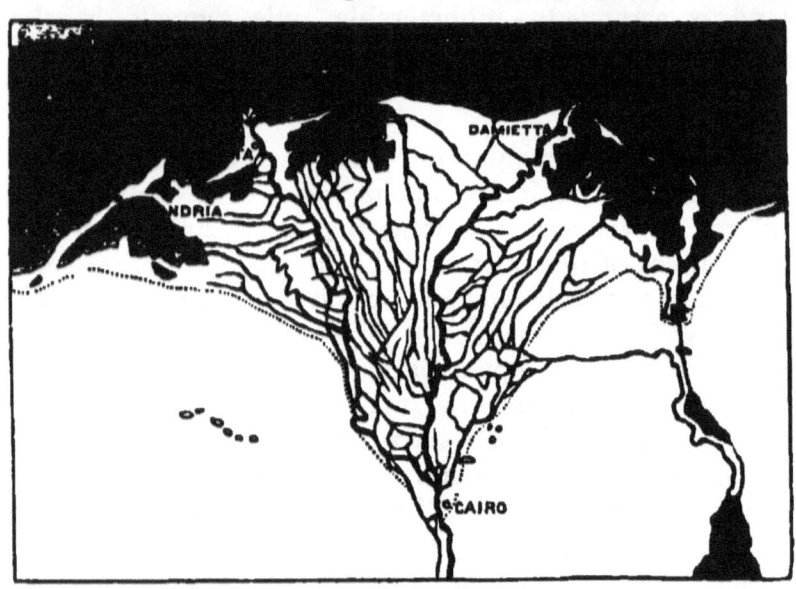

Fig. 52.—Delta of the Nile.

at intervals. Growing, by the addition of many smaller streams from either side, it becomes more and more like a true river, as it rolls along. This valley part is by far the longest and most important in its course. Here it receives most of its water, and puts on its distinguishing characters as a river. Its tributaries are no longer mere torrents or brooks, but rivers, sometimes as large as itself. As these increase in size, however, they become fewer in number, so that in the last stages of its journey, the main stream receives few or no affluents.

6. When at last, quitting the valley which has conducted it through the hills and the lower undulating country, the river reaches the low plains towards the sea, it enters upon the final section of its course, that of the **delta.** Hitherto it has always been receiving, but never giving off branches. But now, reaching the low level land of the delta, it begins to divide, sending off many arms which wind to and fro, and again divide among the swampy flats. So much does the river ramify in this part of its course, that it may enter the sea by many channels, or mouths, often so nearly of the same size, that it may be hard to say which of them should be called the chief. Moreover, these intricate channels are constantly shifting their position as the river-water moves seaward. Hence, what was once the chief mouth is now perhaps half-filled up, and the principal discharge takes place by another channel at some distant part of the delta.

7. It will be noticed on a map, that except where flowing along some straight valley among mountains, most rivers do not run in a straight line for more than a short distance. This is the case in every part of their course. The channel is continually winding from side to side. On a small scale, the same arrangement of curving water-lines may be observed when the rain, during a heavy shower, is running down a sloping piece of roadway. At the upper part, the runnels, though always rushing downwards, cannot, owing to the unevenness of the ground, descend in straight lines, but are deflected, now to one side, now to another, by little pebbles or bits of clay or other roughnesses on the road. Where straight ruts have been made for them by cart wheels, they take advantage of these, and flow then in straight courses, as a river does in a longitudinal valley. But they escape before long, and resume their winding way as before, joining each other in the descent, and swelling the main runnel which sweeps down the road, and eventually finds its way into some neighbouring ditch. What the pebbles, ruts, and other roughnesses on the road are to rain, the uneven surface of the country is to brooks and rivers. In either case, the water seeks the readiest path of escape to lower levels. But this path is seldom the shortest. Every obstacle which the water cannot surmount or remove, serves to turn it aside, and, as such obstacles abound, the flow of the water is a continuous series of turnings and windings. Ridges and hollows, heights and valleys, turn the streams, now to one side, now to another, until, after a journey which may be many times longer than the direct distance from their source, the waters find at last a rest in the great sea. The serpentine curves or meanderings of running water form one of its most characteristic features. They are exhibited by tiny brooks flowing in shallow channels across level meadows, and by mighty rivers that wind in deep gorges over vast table-lands. The curves of a portion of one of the large rivers of the globe are represented in Fig. 53; but a closely similar drawing might be made from the course of many a little runnel or beck.

8. Amidst its bendings, a river will here and there eventually cut through the narrow part of a loop, and thus shorten and straighten its channel. The loop being gradually shut off from the river by an accumulation of sand and mud, becomes a crescent-shaped pool or lagoon of stagnant water. Instances of this kind occur commonly along the courses of streams which flow through flat land, as in the case of the Mississippi (Fig. 53). In the delta of the Rhone they are called " Aigues Mortes," or dead waters.

9. In consequence of the frequent shifting of the direction of a river-course, what is at one place the east side, may be successively the north, south, and west sides in the space of a few miles. It would therefore lead to much confusion were we to speak of the east or west side, or the north or south bank of a river. Accordingly, it

is usual to call one side the **right bank**, and the other the **left bank**, the observer being supposed to be looking down the river in the direction of its flow, so that no matter how much the river may wind, the terms right and left are always correctly descriptive.

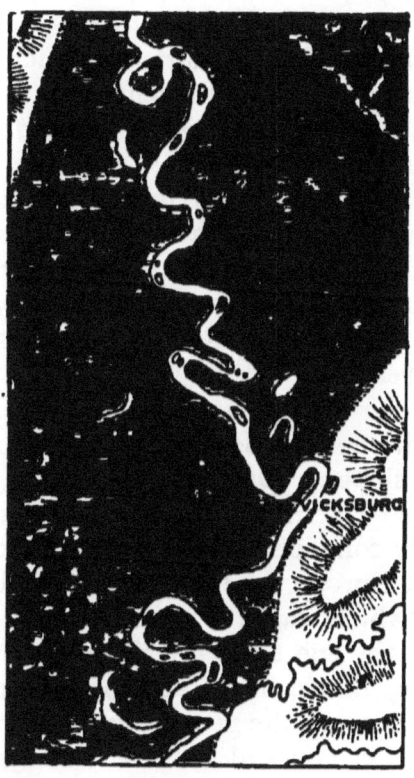

FIG. 53.—Windings of the Mississippi. The shaded part marks the alluvial plain.

10. As may be readily understood from a map, each large river is the natural drain for a wide region. For instance, all the surplus rain and the discharge from melted snow and springs over by far the largest part of North America, find an outlet to the sea by a single river—the Mississippi. The space drained by this river

is computed to be 1,244,000 square miles. This is termed the **drainage-basin** or **catchment-basin** of the river. The drainage-basin of the Ganges is estimated at 432,480 square miles; that of the Rhine at 75,000; that of the Severn 8580; of the Thames at 6960; of the Shannon at 6946; and of the Tay at 2250 square miles.

11. A pencil line traced on the map of North America, round the sources of all the streams which are tributary to the Mississippi, will represent what is called the **water-shed**,[1] **water-parting**, or **divide** of the Mississippi basin. To the north of it lie the basins of the Mackenzie and of the rivers that drain into Hudson's Bay, to the east the St. Lawrence and the smaller rivers of the Eastern States, while to the west the Fraser, Columbia, Sacramento, Colorada, and many lesser rivers carry the drainage of the Rocky Mountain slopes into the Pacific Ocean.

12. The water-shed of a country or continent is thus a line which divides the flow of the brooks and rivers on two opposite slopes. On many maps it is marked as if it were a ridge or mountain-chain. But in reality it does not necessarily coincide with the highest ground. Trace on the map of Europe, for example, a pencil line between the streams which drain to the Atlantic, Baltic, and North Sea on the one side, and those which drain to the Mediterranean, Black, and Caspian Seas on the other. Such a line will mark the general water-shed or divide of the continent, but you will observe that the line, instead of running along, and coinciding with, any central mountain-chain, crosses all the great mountains, table-lands, and plains. Beginning at Gibraltar, it traverses the table-

---

[1] Sir John Herschel proposed to write this word *water-sched*, meaning "separation of the waters, not water-*shed*, the slope *down which* the waters run" (*Physical Geography*, p. 120). But the original meaning of "shed" was to divide, or part, and this use of the word still holds in Scotland, where a girl is said to *shed* her hair in the middle. The idea of disjunction is a secondary one, which has gradually come to be the common usage of the word.

land of the Spanish Peninsula, crosses obliquely the chain of the Pyrenees, passes athwart the plateau of Central France, on the right bank of the Rhone, runs through and across the ranges of the Alps, the Black Forest, and the Carpathians, and then descends into the vast plains of Russia, across which it winds in a general north-easterly course to the chain of the Urals. That the water-shed need not be a high ridge, may often be noticed even on low or gently undulating ground, where the same valley may have a stream at either end, flowing in opposite directions, the water-shed being a quite imperceptible rise of the surface between them.

13. Some important lessons as to the form of a country may be learnt by noting the line of its water-shed upon a map. That line seldom runs down the centre of a region like the ridge on the roof of a house. Very commonly it lies much to one side, and winds in great curves as it traverses the land. Now the position of the water-shed, like the axis described in Lesson XIX. Art. 25, suffices to indicate the relative slopes of the two sides of a continent or country, especially where these two sides descend to the sea-level. Take South America by way of illustration. The water-shed of the whole continent lies near the western coast-line. The slope facing the Pacific must thus be very much steeper than that which looks towards the Atlantic. A heavy shower of rain falling on the water-shed will of course run off partly to the west and partly to the east. The westward portion, starting from a height of perhaps 10,000 feet, will reach the Pacific after a journey of not more than seventy miles in direct distance; while the other half, setting out from the same elevation, will have a journey of about 2000 miles in a straight line before it can enter the Atlantic. The water-shed of Hindostan, south of the Gulf of Cambay, is another example of this one-sided position. On a smaller scale, Scandinavia, Great Britain, and Spain illustrate the same feature.

14. When a water-shed runs close to one edge of a continent, there is no room for large rivers on that side; these must flow on the opposite slope. America again furnishes an admirable instance of this obvious arrangement. In South America, for example, there is no river of large size flowing down the short slope into the Pacific, but on the east side, the largest rivers of the world bear the drainage of the continent into the Atlantic.

15. **Sources of Rivers.**—Every shower of rain that falls, and every spring that rises, within the drainage basin of a river, may be regarded as one of the sources of the river. In many parts of the world indeed, such, for example, as the central and southern regions of India, where there are dry and rainy seasons, and where the rivers do not take their rise in high snowy mountainous ground, the water that floods the streams during the wet months is mostly derived from the rain which runs off the saturated soil. In common language, however, the source of a river is understood to mean the point from which the head-waters of the main branch of the river take their rise. It is often hard to say which of the branches of a large river should be called the chief. One may be largest in volume of water, another greatest in length of course. One of the branches has usually been selected by the people of the country, and called by them the main stream.

16. Large rivers rise from various sources—springs, rains, melted snows, or the ends of glaciers (Lesson XXVIII.) and lakes. A great proportion of them may be traced up till their farthest little tributary brook is found gushing out as a spring from the side of some hill or mountain. In limestone countries, as stated in the foregoing Lesson, large rivers sometimes issue from the caverns by which the underground rocks are there perforated. Occasional or periodical rain directly supplies much of the water of many rivers. The Nile, for example, owes its annual rise to the heavy rainfall of the wet

season among the mountains of Abyssinia. The snow-fields of the higher mountains furnish unfailing nourishment to many of the largest rivers of the globe. Thus in Europe, the Rhine and the Rhone take their rise from the melted snow and ice of the Alps. In Asia, the rivers of Northern India descend from the snows and glaciers of the giant chain of the Himalaya. In North America, the abundant patches of snow which mark the higher parts of the Rocky Mountains supply part of the water which drains from the west into the great valley of the Mississippi. Here and there, among the higher summits of the land, little hollows arrest the first runnels of melted snow or of springs, and form little lakes, out of which the infant waters of important rivers flow. Or a lake on lower ground at the confluence of several tributaries fills a vast hollow of the land, whence the united drainage escapes by one large river, like the Rhine from the Lake of Constance, or the Rhone from the Lake of Geneva.

17. **Proportion of the Rain-fall carried to sea by Rivers.**—Since the size of rivers depends upon the amount of rain-fall or snow-fall within their drainage-basins, we may naturally inquire how much of the total moisture discharged from the air upon the land is actually returned to the sea by rivers. The proportion between rain-fall and river-discharge has never been very satisfactorily determined, but is said to vary from $\frac{1}{3}$ to $\frac{1}{4}$; that is to say, only about a third or a fourth part of the water which falls upon the land as rain or snow is carried off by streams. The greater portion is returned to the air again by evaporation. From the moistened soil, from every surface of snow or water, from each spring, brook, lake, and river, vapour is continually passing into the air. The rivers, therefore, do not bear to the sea even all the water poured into them, for they are continually losing water by evaporation from their surface.

18. The diminution of rivers, the drying up of brooks, and the cessation of springs, during seasons of drought, show how dependent is the flow of water over the land upon its circulation through the air. In countries like Britain, where heavy rains may occur at any time of the year, the rivers are subject to irregular increase. In those regions, however, where a wet and dry season succeed each other at certain intervals, the rivers have their periodical rise and fall. The most familiar example of this regularity is that shown by that historical river—the Nile. Egypt, through which this river flows, is a singularly dry country, wherein rain seldom falls, and yet every year, and with such regularity that almost the very day of the change may be foretold, the river begins to rise, and continues to do so until the low plains on either side are overflowed. It then slowly subsides, leaving a film of fine mud over the ground, and resumes its former channel. This remarkable feature greatly puzzled the ancients. They accounted for it by supposing that the Nile rose among snowy mountains far to the south, and that the inundations were caused by the melting of the snows. But in recent years the true cause has been ascertained. The high and rocky table-land of Abyssinia is visited by heavy rains during the months of March and April. The numerous gullies and gorges which intersect that rugged country, and which were previously quite dry, are then filled with torrents of water, which rush down and swell that branch of the great river known as the Blue Nile. It is these annual rains in the far highlands of Abyssinia, therefore, which cause the regular inundation of the rainless plains of Egypt.

19. In countries liable to heavy periodical rains, the relation between the flow of rivers and the rain-fall is made strikingly clear. The Mahanadi River in Central India, for instance, has its basin within the area to which the south-west monsoon brings the copious rains alluded to in Lesson X. Falling upon high rocky ground, the rain at once rushes down by innumerable channels to swell the main river, which, unable to find in its channel room for all this water, overflows the surrounding country, producing great havoc in its course. But in the dry season, from February to May, the river is low, because, as little or no rain then falls, it depends for its water upon the supplies which it receives from springs.

20. Another illustration of the relation between rainfall and rivers is supplied by those singular dry ravines and gravel-tracks, so common in Syria, Arabia, and the Valley of the Euphrates, called "wadys." Those which lie within tracts where there is a rainy season, are turned into water-courses during the wet part of the year. But over a great part of the region where they occur, little or no rain falls, so that they remain constantly dry. That they were once the channels of brooks and rivers cannot be doubted. It is apparent that a change has taken place in the climate of these districts, perhaps partly caused or at least aggravated by the destruction of the ancient woods and forests, and the abandonment of the cultivation of the soil. The rains that once refreshed the land have ceased to fall, the river-beds which carried the surplus water to the sea are now dry, and the valleys are parched and barren.

21. In a country which is subject to heavy rains at different and uncertain times of the year, the rivers are liable to be swollen on any day; and as heavy rain often succeeds dry weather, the rivers may pass rapidly from a state of low water to full flood. Where, as in the case of the Nile, the rain comes regularly at the same season, the river swells and falls again with remarkable slowness and uniformity. But there is another cause of the regular, as well as irregular, flooding of rivers. Where a river draws its supplies of water in great part from the melting of snow among the mountains, it will evidently have a larger volume in summer and autumn than in winter and spring. The Rhine and Rhone, for example, which take

their rise among the snows and glaciers of the Alps, fill their channels with water during the dry hot weather of July and August, and shrink in size during the cold and often wet months of the year. Besides this annual increase and diminution, the upper course and higher or mountain tributaries of such snow-fed rivers are liable to occasional and sometimes disastrous floods in summer, not caused by heavy rains, but by dry and warm weather. When the warm south wind called *Föhn* blows upon the snowy slopes of the Alps, it causes a rapid thaw. Torrents of milky snow-water pour down the sides of the mountains and feed the brooks, these soon swell and rush forward into the rivers, which rapidly rise and sweep down their valleys, often carrying away bridges, inundating meadows and strewing them over with gravel, or even breaking down parts of villages and towns built upon their banks.

22. Flow of Rivers.—Standing by the side of a broad rushing river, trying to estimate the rate at which the water is moving, and how much passes us in an hour or day, we should probably, in most cases, guess the rate of motion to be faster than it is. The broad stream, with its eddies and gurglings of water, impresses us with the idea of rapidity as well as of volume. The rate of flow is determined mainly by the angle of slope, partly also by the volume of water, and the form of the channel. The average angle of slope of river-channels is considerably less than we might have expected it to be, that of the larger rivers of the continents probably not exceeding two feet in the mile. The Missouri has a mean fall of 28 inches in the mile; but the Volga does not slope more than 3 inches in the mile. To be easily navigable, a river should not have a mean declivity of more than 10 inches in the mile or 1 in 6336. Of course, in their mountain-track, streams have much higher angles of descent. That of the Arve at Chamounix is 1 in 616, and that of the Durance varies from 1 in 467 to 1 in

208. But such rapid declivities are those of mountain torrents rather than of rivers.

23. A moderate rate of the flow of a river is about $1\frac{1}{4}$ mile in the hour; that of a rapid torrent does not exceed 18 or 20 miles in the hour. The larger rivers of Britain have a velocity varying from about 1 mile to about 3 miles in the hour. Hence we may walk by the margin of a river and easily outstrip the rate of its current.

24. But the water of a river does not flow in every part of the channel at the same rate. Owing to friction against its bed, the river moves slower at the sides and bottom than in the middle. On a small stream, we can easily prove this by throwing in pieces of wood, and watching how much faster those travel which fall in the middle, than those which have lighted near the edge. Evidently, therefore, the addition of more water to the same channel will increase the rate of motion of the stream. When two rivers join into one, the united stream may occupy a channel no broader than one of them did before, but, without acquiring any increased slope, it runs faster, because the water has now to overcome the friction of only one channel instead of two. For the same reason, a river confined in a deep narrow gorge runs faster than where, with a like declivity, it spreads over a broad gravelly bed. When this cause is kept in mind, we can understand the meaning of the curious fact that a river is sometimes not increased in breadth, even by the influx of large tributaries. Thus the Mississippi becomes no wider, even after it receives its largest affluents. Eighteen hundred miles above its mouth it is 5000 feet, or nearly a mile, broad. But at New Orleans, where it enters the Gulf after having been successively fed by the vast streams of the Missouri, Ohio, Arkansas, and Red Rivers, it is only 2470 feet broad, or less than half a mile, and yet in that comparatively narrow channel the drainage of nearly half a continent passes out to the sea.

25. Volume of Water discharged by Rivers.— Measurements and estimates have been made of the amount of water carried annually into the sea, or hourly past certain places by different rivers. The Mississippi, for example, has been found, after a careful survey of its operations, to discharge annually into the Gulf of Mexico no less than 21,300,000,000,000 cubic feet of water, enough to make a lake as large as the whole of England and Wales and twelve feet deep. The Danube sends into the Black Sea 207,000 cubic feet of water every second. The Tay, though not the largest river in the British Islands, brings down more water into the sea than any other, its annual discharge being estimated at 144,020,000,000 cubic feet.

Lesson XXVI.—*The Waters of the Land—Part III.*

Lakes and Inland Seas.

1. Owing to inequalities of the surface, the water that falls upon the land cannot always flow off at once into the sea. Hollows occur which intercept the drainage, until it fills them and flows over from the lowest parts of their margins, or otherwise escapes. Such water-filled basins are called **lakes**, but when they are of large size and are occupied by salt water, they receive the name of **inland seas**.

Lakes.

2. At first lakes might be expected to lie indifferently on any part of the earth's surface. Any good general map of the world, however, shows that they are not distributed altogether at random, but are chiefly developed in certain regions. An attentive study of this development throws some light upon the problem of the origin

XXVI.]   LAKES AND INLAND SEAS.   259

of the hollows, and why they should be where we now find them.

3. (1) First, then, in the northern hemisphere, lakes may be observed to be scattered in extraordinary abundance over the northern parts of Europe, down to about

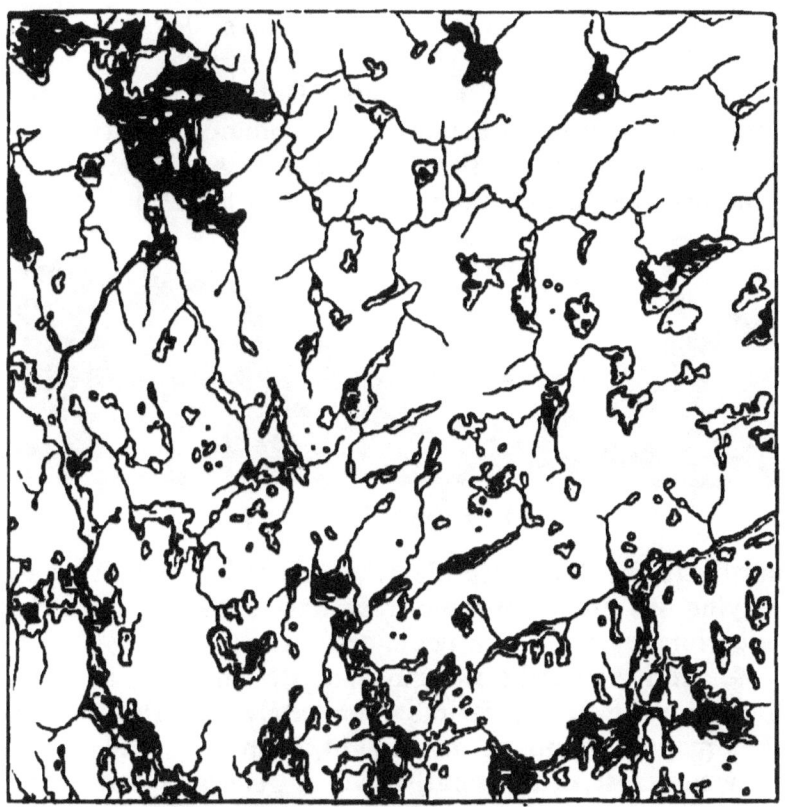

FIG. 54.—Part of the Island of Lewis, illustrating the abundant lakes of the north-west of Europe.

the 52d parallel of latitude, and in America to about the 42d parallel. In some districts within these regions, there seems to be almost as much water as land. In Finland, for example, nearly a third of the country is covered with lakes and marshes, and as there are no mountain-ridges, and no dominant lines of valley, the undulating surface looks like that of a land which has been half submerged. In the north-west of Scotland similar scenery occurs; from some of the hill-tops there, the undulating surface of the low country is seen to be plentifully strewn with lakes. In the accompanying figure (Fig. 54) a representation, on the scale of one inch to a mile, is given of a portion of the surface of Lewis, one of the islands of the Outer Hebrides, from which some idea may be formed of the abundance and irregular distribution of these sheets of water. Again in North America, the British possessions and a large part of the

Fig. 55.—Section of a lake-basin excavated in solid rock.

north-eastern States are plentifully dotted over with lakes, varying in size from such vast basins as that of Lake Superior down to mere pools or tarns.

4. In these northern regions, as the lakes are not confined to valleys, but lie indiscriminately over the surface, they may be found at any height, from the sea-level to near the crest of the land, which is generally undulating and hilly rather than mountainous. The abundance of these lakes has been associated by many geologists with the fact that the land in the northern hemisphere has been covered with sheets of ice which, grinding down the surface, have left it strewn with clay, gravel, and sand (Lesson XXVIII.). Many of the lakes lie in hollows excavated out of the solid rock (Fig. 55), which still retains abundantly the scratches and groovings made

XXVI.]   LAKES AND INLAND SEAS.   261

upon it by the movements of the ice. Other sheets of water are enclosed among heaps of debris, left on the land when the ice melted (Fig. 56).

FIG. 56.—Section of a lake-basin lying in a hollow of superficial detritus.

5. (2) A further feature in the distribution of lakes, which may be observed on maps, is the more or less abundant occurrence of these sheets of water among mountains. Take Europe as an illustration. Even in the comparatively low mountain groups of Scotland, Cumberland, and Wales, lakes abound, forming one of the great charms of the well-known scenery of these districts. Among the Alps a series of large lakes occurs on each side of the main axis of the chain, and innumerable minor sheets of water

FIG. 57.—Section of a lake dammed up by a barrier of earth or gravel.

occur scattered at all heights among the central mountains, up even to the borders of the snow-line. All mountain systems, indeed, have not the same abundance of water-filled hollows; in some there are few or none. Lakes among mountains may be, in some cases, hollows formed during the elevation of the mountains (Lesson XXIX.); in other examples, like those referred to in Art. 4, they may either have had their basins scooped out by glaciers or formed by the irregular piling up of ice-borne debris (Fig. 57). In most volcanic districts, lakes occur in cavities which have been formerly blown open by explosions from below.

6. (3) A third series of lakes may be observed to occupy depressions on table-lands. The most remarkable examples are furnished by the great lakes of Equatorial Africa. Of these the Victoria Nyanza lies at a height of about 3300 feet above the sea, and is believed to cover an area of not less than 30,000 square miles. On the same continent a vast depression, with many small lakes, extends westward from the Nile valley, and stretches between the southern limits of Barbary and the country drained by the River Niger. Another smaller hollow lies in the southern part of the continent, and contains some small lakes, of which Lake Ngami (2900 feet above the sea) is the largest. On the great table-land of Asia numerous lakes occur over Thibet, Turkestan, and Mongolia.

7. It will be noticed that many of these table-land lakes have no outlet, but lie in hollows below the general level of the surrounding country. They receive supplies of water from the streams which drain into them, but no river escapes even from the lowest parts of their borders. Now almost all lakes of this kind are filled with salt water. That they should be salt, will be found on reflection to be a necessary consequence of the solvent action of running water on the land (Lesson XXIV. Arts. 22-36). Water flowing over or through rocks dissolves out of them some of their soluble ingredients. Among the substances so removed, common salt, sulphate and carbonate of lime, and sulphate of magnesia, are of common occurrence. These dissolved salts are carried down by rivers, and in most cases find their way ultimately into the sea. But in the depressions of the great table-lands, the hollows into which the water drains and finds no outlet may be compared to great evaporating vats or troughs, like those in which sea-water

is boiled down in the manufacture of salt. Water is constantly poured into them and none flows out; yet their level is not rising, because, as fast as the water enters, it passes off again in invisible vapour into the atmosphere. The supply of water and atmospheric evaporation just balance each other, and any cause that diminishes the one or the other would make the level of the lakes rise or fall. But the vapour which rises from the inland basins leaves behind in the water the various saline solutions carried into them by rivers. Hence year by year these lakes and pools become salter. When they dry up, they leave a crust of salt upon the ground which they once covered. The soil is in such places so impregnated with salt that plants will not grow upon it, and its arid sandy surface stretches for leagues as an inhospitable desert.

8. Where a fresh-water lake does occur in these regions it will be found to have some outlet by which its surplus water is removed, so as to prevent increase of saltness. Lake Chad in Central Africa, for example, lies in a hollow from which no river escapes to the sea, and it was believed in consequence that this fresh-water lake had no outflow, and was thus an exception to the general rule. More recently, however, a river has been found opening from its north-eastern margin and carrying the overflow along a wide valley, in which the water is finally dried up amid sandy wastes.

9. In the eastern countries bordering the Mediterranean many salt-lakes occur, as well as ground incrusted or impregnated with salt. The naturally-formed salt has been used from time immemorial by the dwellers there. A little reflection on the mode of its formation and its composition explains the meaning of a curious passage in the Bible which is not in itself very intelligible. It occurs in the Sermon on the Mount. "If the salt have lost his savour, wherewith shall it be salted? it is thenceforth good for nothing, but to be cast out, and to be trodden under foot of men." (St. Matthew v. 13.)

Were the substance here spoken of our common salt (chloride of sodium), it would be difficult to explain how it could possibly lose its taste without ceasing to exist at all. Its taste is to us as essential a character as its chemical constitution. But no doubt the substance referred to was the white incrustation obtained in the East from the sides of salt lakes and the bottoms of dried-up saline pools. Now this incrustation, besides common salt, contains also carbonates or sulphates of magnesia, lime, soda, and other ingredients. Of these various components, common salt is one of the most soluble (Lesson XIII. Art. 5), that is to say, it is among the last ingredients to appear when the water is evaporated, and among the first to disappear when moisture is supplied again. We can see, therefore, that the white natural crust used by the people of Syria for salt, being kept for a time and exposed to damp or rain, might lose all its salt. The more insoluble residue, consisting of gypsum, carbonate of lime, etc., though in appearance unchanged, yet having little or no taste, would be quite useless for the purposes for which the salt had been gathered. The question, therefore, might well be asked—" If the salt have lost his savour (or taste), wherewith shall it be salted?"

10. (4) A fourth series of lakes occurs along many parts of the margin of the land where the ground is low, and consists of soft sandy, clayey, or gravelly materials. These maritime sheets of water are known by the name of **lagoons**. In Europe they fringe all the Prussian shores of the Baltic, reappear on the west of Denmark, Holland, and Belgium, and are found at intervals along the northern coasts of the Mediterranean Sea, from the east of Spain to the western shores of Greece. In Asia they extend for hundreds of miles along the eastern and western sides of the peninsula of Hindostan. In America a long line of them skirts the Atlantic sea-board of the United States.

11. Lagoons along the sea-margin are, for the most

part, shallow and narrow, running parallel with the coast, from which they are separated by a strip of low land formed of sand, gravel, or other loose materials. When the sea flows into them, the waters are salt or brackish. When they communicate with the sea only by a narrow outlet, or when they have no outflow, but soak through a porous bar which banks them out from the sea, they are fresh.

12. Most lakes derive their water from streams that flow into them. In Fig. 54, for example, this connection is well shown, though there the feeders of the smaller lakes are too minute to be shown upon the map. Many valleys contain chains of lakes along their course. A river flowing in one of these valleys, appears alternately to contract and expand, having a comparatively rapid flow where it takes its own river-form, and losing itself in still water when it enters a lake. It would seem that this arrangement was formerly commoner than now, for in many mountainous districts the rivers wind to and fro across flat meadows where no doubt lakes once lay. The meadows are, in fact, the alluvial plains formed by the gradual filling up of the lakes, owing to the deposit of sediment brought down by the inflowing streams (Lesson XXVII. Art. 30).

13. But lakes likewise derive their supplies partly from springs which rise beneath them. In some cases, indeed, the whole supply comes from such underground sources. The Lake Zirknitz near Trieste (Lesson XXIV. Art. 42) affords an excellent illustration of this feature, for it comes and goes with the seasons. After long drought it disappears. When heavy rains fall on the surrounding mountains of Carniola, which are formed of remarkably honeycombed limestones, the water at first sinks out of sight, but after filling the underground passages, it comes out with a roaring noise from the funnels and caverns which open upward into the hollow of Zirknitz, which is then converted into a lake. No stream flows out of it, since both its supply and its overflow pass away by underground channels. This example shows how lakes, like rivers, depend ultimately upon the rainfall for their water, and are apt to vary in level with the wetness or dryness of the seasons. In Northern Africa also, the Sebka-el-Faroon, a hollow 100 miles in length, lying to the South of Tunis, at a level of several feet below the Mediterranean, is in winter covered with water to the depth of two or three feet. Having no outlet, and undergoing rapid evaporation during the parching summer of that hot climate, the lake disappears and leaves a salt-crusted floor.

14. Some of the largest fresh-water lakes in the world are those of North America, Lake Superior alone covering an area of 23,000 square miles, its surface being 627 feet above the level of the Atlantic, and its average depth nearly 1000 feet. Another series of vast sheets of fresh-water, already (Art. 6) referred to as lying on the tableland of equatorial Africa, forms the source whence the Nile and the Congo take their rise. In the heart of Asia, Lake Baikal stands at a height of 1363 feet above the sea, and covers a space 370 miles in length by from 20 to 70 in breadth. On these vast sheets of water storms arise on a scale hardly inferior to those on the sea itself. Fresh-water, being lighter, is more easily stirred by the wind than salt-water. It is soon raised into ripples, and when deep and wide, and driven onward under the pressure of a continuous high gale, it rises into large waves, which roll across and burst in foam against the windward shore, heaping up gravel and sand, or cutting down the cliffs, as is elsewhere done by the waves of the sea (Lesson XVIII. Art. 11).

15. The depth of lakes varies almost within as wide and indefinite limits as their size. The form of the land round a lake generally affords some indication of the probable depth of the basin, as the depth of the sea

near the land may be inferred from the contours of the shores. Thus, if the ground is low and slopes gently into the water, we may infer with some confidence that the lake must be shallow. If, on the other hand, the ground rises steeply out of the water, as when a lake fills a valley between two precipitous mountain-ridges, we may be prepared to find that the lake is deep. Several of the Alpine lakes attain a great depth. Thus Lake Como is nearly 2000 feet deep, and the Lago Maggiore 2800 feet. In both of these lakes the bottom sinks below the level of the sea, that of the Lago Maggiore being 2149 and that of Como 1318 feet below the surface of the Mediterranean. But they all have outlets, and their waters are fresh. In Scotland, Loch Ness forms a remarkable feature in the great valley which cuts the Highlands in two, its surface being about 70 feet above the sea, and its greatest depth 810 feet. The bottom is therefore not only below the sea-level, but actually a good deal deeper than any part of the North Sea between Scotland and Denmark.

16. Little has yet been done in exploring the depths of these profound hollows on the land. Some of them, like Loch Ness, were at a comparatively recent period filled by the sea, and though probably all trace of the salt-water has long since been removed, perhaps some lingering forms of life may have survived the change, and, though possibly somewhat modified by the changed conditions in which they have lived, may exist still in the undisturbed abysses, somewhat in the same way that certain common shore-plants, such as the coast-plantain (*Plantago maritima*), are met with on the tops of the Scottish mountains.

17. It would be interesting also to know the distribution of temperature in the water of these deep lakes. According to observations made in Loch Lomond, in Scotland, the depth of which is about 600 feet and its surface 25 feet above the sea, it appears that a tolerably

constant temperature of about 42° Fahr. characterises the lowest stratum of water for 100 feet or so above the bottom. The cold water of winter must sink down, and as the sun's rays can sensibly heat the water only for a short way beneath the surface, the temperature of the deeper parts of such lakes must be kept permanently low. In the Lake of Geneva in autumn, while the surface water shows a temperature of 78° Fahr., the bottom water, at a depth of 950 feet, marks 41°·7. Similar observations on the other deep lakes of Switzerland and Northern Italy show that they all have a permanent mass of cold water at the bottom. Farther south the Lago Sabatino, near Rome, was found to have a temperature of 77° at the surface, but one of 44° at a depth of 490 feet.

18. One great and useful office of lakes is to exercise an important equalising influence on temperature, preventing the air around them from being so much heated in summer and so much cooled in winter as would otherwise happen. (Lesson XXXI. Art. 20.) The mean annual temperature of the surface water of the Lake of Geneva as it issues into the Rhone is nearly 4° warmer than that of the air. A second and still more important function of lakes is to regulate the flow of the rivers that issue from them. They receive the water discharged by heavy rains and rapidly-melted snows, spread it over a large surface, and allow it gradually to escape by the outflowing river. In this way they prevent the occurrence of those sudden and destructive floods, which, in the absence of such natural reservoirs, are apt to occur in all countries subject to copious rain, or where large masses of snow may be quickly thawed. A third part played by lakes is that of arresting the gravel, sand, and mud brought down by the streams which flow into them. These materials, which so often discolour the tributary brooks and rivers, fall to the bottom when the currents are checked by entering the still lake-water. Lakes in this way filter

the rivers. The Rhone, for instance, is a muddy stream where it enters the Lake of Geneva, but at the lower end, where it quits the lake, its water is as pure and limpid as that of a spring. Its sediment has been dropped upon the bottom of the lake, which must consequently be slowly rising. The Lake of Geneva is being gradually filled up in the manner already referred to in Art. 12.

Inland Seas.

19. From what has been said in Arts. 6-9 the learner will have no difficulty in tracing out upon a map those areas of the earth's surface where the lakes must be salt. In every tract of land into which rivers flow without an outlet, and where the surplus water passes off by evaporation, brackish or salt lakes may be looked for. A salt lake need not necessarily have been once connected with the main ocean. So constantly are salts present in fresh-water that any fresh-water lake, where the only escape for the water is by evaporation, will eventually become salt. The salt lakes on the table-lands of Asia and Africa were, no doubt, fresh at first, and have gradually grown salt as the process of evaporation has been continued century after century. In North America, the Great Salt Lake at Utah, lying at a height of 4200 feet above the sea, and covering a space of 75 miles in length by from 15 to 40 in breadth, together with many other smaller salt lakes in the same region, afford most interesting evidence of the process whereby fresh-water lakes, by change of climate, leading to diminished rainfall and increased evaporation, become saline and even intensely bitter.

20. But in some parts of the world there exist sheets of salt-water which are either still connected, or can be shown to have been once connected, with the main body of the ocean, from which they have been separated by

subterranean movements. **Inland seas,** now completely isolated, may have their surface below the level of the main ocean, or they may have been carried up together with the land around them, so as to lie above that level. By far the most remarkable of these sheets of water is the chain of inland seas and salt lakes which extends from the Black Sea and Sea of Azov eastward into the basin of the Caspian and Aral Sea. At one time the Arctic Ocean extended for some way southward across what are now the "tundras" or frozen plains of Siberia, while the Black, Caspian, and Aral Seas formed a united mediterranean sea, though it is not ascertained that this sea was once united to the Arctic Ocean. Europe and Asia were thus, in some measure at least, separated from each other by arms of the sea. Owing to underground movements this separation has been in great measure obliterated. But in the deeper cavities of the ancient depression portions of the sea still remain, retaining even yet the marine shells, fishes, and seals which abounded in the water before the elevation of its bed. Of these relics the largest and most important is the Caspian Sea, which lies 84 feet below the level of the Black Sea, is from 2000 to 3000 feet deep in the central parts, and covers an area of about 180,000 square miles. It receives the drainage of the whole of the south-east of Russia in Europe by such important rivers as the Volga and the Ural, but it has no outlet. So large is the mass of freshwater poured into the Caspian, that the saltness of the greater part of that sea is not more than about one-third that of ordinary sea-water (Lesson XIII. Art. 6). But along the shore numerous lagoons occur, where, in the dry and hot weather of summer, so much evaporation goes on that the water becomes intensely bitter and salt, and saline incrustations form at the bottom and on the shores. On the east side of the sea, lies the wide but shallow Karaboghas Bay, which may be looked upon as a vast evaporating basin. A current is always passing in through the narrow opening, but there is said to be no compensating under-current outwards, so that, as the water-level does not rise, all this constant inflow must be supplying the loss by evaporation. The bay, therefore, grows every year salter. A solid layer of salt forms on the bottom of some of the shallower parts of its shores, and the water is there so saline that a cord on being let down into it and pulled up again is immediately crusted over with salt. Seals, which used to flourish there, are said to have been driven away by the increasing saltness of the water.

21. The Sea of Aral fills another of the hollows in the same ancient depression between the European and Asiatic high-grounds. It is a lake of brackish water 265 miles long and 145 broad. It is said to be at a height of only 33 feet above the level of the sea. On its southern side it receives the Oxus, which carries into it the drainage from the northern slopes of the great chain of the Hindu Kush mountains. It likewise obtains supplies of water and mud from the Jaxartes, which takes its rise among the lofty Thian-Shan mountains. Yet the Sea of Aral, like the Caspian, has no river flowing out of it. It loses by evaporation as much water as it receives. Indeed, the loss from this cause would seem at present to be greater than the supply of water, for the sea is said to be sensibly decreasing in size.

22. The valley of the Dead Sea is remarkable as being the most depressed on any part of the land of the globe. The surface of that sheet of water is 1298 feet below the level of the Mediterranean Sea. The water, so intensely salt as to be a kind of brine, contains in every 100 parts rather more than 24 parts by weight of salts, or about seven times the proportion in ordinary sea-water.

LESSON XXVII.—*The Waters of the Land—Part IV.*

The Work of Running Water.

1. In the two previous Lessons we have followed the circulation of running water over the surface of the land in an elaborate network of branching water-courses, which, stretching from the slopes of the central mountains down to the sea-shore, carry back to the ocean the surplus drainage of the land. We have seen how in the hollows of the land the water gathers into lakes, yet that it does not accumulate indefinitely there, since it either overflows and again takes the form of streams, or passes off by evaporation. Delivered to the sea, and mingling once more with that great body of water, it is anew raised by the sun's heat into invisible vapour, and carried by winds across the land to begin again the same circulation.

2. So vast a body of water, ceaselessly moving over the land, slowly, but in the end, extensively, modifies the forms of mountains, hills, valleys, and plains on which it falls and flows. We shall now consider the nature of the change thus effected, beginning with the tiny raindrop, and tracing the operations of running water down to the mouths of the great rivers as they enter the sea.

3. **Rain.**—The action of rain in washing the air was described in Lesson X. Art. 35. Its further influence in decomposing rocks underneath the surface of the ground was traced in Lesson XXIV. Similar changes take place upon the surface of the land. Rain-water, by means of the carbonic acid which it takes out of the air and soil, or the organic acids which it absorbs from decomposing vegetable matter, attacks rocks exposed to the air, dissolving and removing the more soluble parts of them, thereby loosening their cohesion and causing them gradually to crumble down. Calcareous rocks, like marble, suffer much from this kind of waste; but even

XXVII.] THE WORK OF RUNNING WATER. 273

hard rocks like granite do not escape. The solvent action of the acids upon some of the ingredients of the stone is a chemical process. But when the outer layer or crust of rock has, in this manner, been loosened, heavy rain may wash off the disintegrated particles, and thereby expose a new surface to further decay. The action now becomes mechanical. These two combined processes powerfully influence the scenery of the land (Lesson XXIX.).

4. The little prints which the rain-drops leave upon a surface of moist clay or sand offer the simplest instance of the mechanical action of rain (Fig. 58). From such apparently trivial effects many stages may be traced, until

FIG. 58.—Prints made in soft mud or moist sand by rain-drops.

we reach huge pillars, like those shown in Fig. 59, which have been carved out by the blows of innumerable rain-drops. The material of these columns is a stiff earth or clay, stuck full of stones and large blocks of rock, and readily crumbling down under the influence of the weather. The large blocks, remaining of course unwasted, serve each to protect the portion of the earth lying underneath it, while the surrounding clay is washed down. Thus the block becomes, as it were, the capital of a pillar which seems to rise slowly out of the rest of the earthy mass. Each pillar stands as a monument of the continuous waste, somewhat in the same way as the columns of rock or clay, left by the workmen in a railway-cutting or quarry, show the extent of material which has been removed. On reaching the more level ground, the runnels of rain drop the particles of earth, washed by them

T

from the slopes and steep faces, and spread them out over the soil.

5. Here then, in the action of the rain-drops, a sort

FIG. 59.—Earth-pillars of the Tyrol (from a photograph).

of type may be seen of the work of all the great rivers of the globe. It is threefold. First we have Erosion, in the loosening of the particles of earth or rock,

Secondly, Transport, in the removal of these loosened particles, and the consequent exposure of fresh surfaces to waste; and thirdly, Deposit, in the laying down of a new stratum formed out of the removed materials. Let us now see how these three kinds of action are manifested over the surface of the land by rivers.

6. **River-Erosion.**—Every runnel, brook, and river, in short every current of water, no matter how small, which moves over the land is busy removing part of the soil or rock over which it flows. This work, like that of the rain-drops, is twofold. In the first place, the rain and the water of the streams, dissolving certain parts of the solid substance of the land, and carrying them away in chemical solution, effect a considerable amount of waste in countries where the rocks consist of limestone, or contain a marked proportion of soluble substances, although the running water is not visibly affected in colour or transparency. Some idea of the extent of the loss thus sustained by the surface of the land may be formed from the amount of dissolved mineral matter which is found in the water of rivers (Art. 14).

7. In the second place, the solvent action of rain and the disintegrating effects of frost (Lesson XXVIII.) cause the surface of exposed rocks to crumble down into loose clay, sand, and angular rubbish, even the hardest rocks being split into fragments. This debris of the land, washed away by the rain and brooks, becomes the instrument of still further destruction. As it is hurried along, its particles, ground against each other, are reduced still further in size, and at the same time are worn smooth and round. They thus acquire that familiar **water-worn** character which we recognise as the most obvious and distinguishing feature of the detritus in the channel of a stream. So constant is this character, that when, at any point in the course of a river, sharp-edged fragments appear on the banks, we naturally conclude that they cannot have lain very long in the current, nor have travelled very far.

The fragments, the longer they are rolled about, and the farther they are carried, grow smaller in size, until at the far end of the river they may be found as mere fine sand and mud. The source of this fine sediment must be sought among the rocks of the far distant mountains, in the higher part of the river-basin. In many rivers it may be traced upward through every gradation of sand, gravel, shingle, and boulders, until its origin is found in the huge blocks and abundant angular rubbish loosened from the parent cliffs, which in the course of ages have supplied a constant and abundant tribute of detritus to the river. In some respects a river may be compared to an enormous grinding-mill, where large pieces of stone go in at one end, and only fine sand and mud are seen to emerge at the other.

8. But the loose materials swept away by the streams not only wear each other down, they likewise erode the sides and bottom of the water-courses. The water-worn character is thus not confined to the loose sand, gravel, and stones, but is as marked upon the solid rocks over which these materials are driven. Even the hardest kinds of stone cannot resist constant friction. They become smooth and polished, though, where out of the scour of the water, they may present a rough surface and sharp edges. The upper limit of the grinding action of the flooded stream is in this way well defined along the sides of a rocky ravine.

9. In the course of time the stream grinds out a channel for itself through even the hardest rocks. This channel, however, is seldom a mere deep straight trench. Since rocks offer many varying degrees of resistance to erosion, they are worn down unequally, being scooped out where more easily worn away, and left projecting where more durable. The stream, thrown from side to side, dashes along, sweeping onward the sand and mud which it drives over its rocky bed, and excavating those winding picturesque ravines which are such familiar features in water-courses.

XXVII.] THE WORK OF RUNNING WATER. 277

10. Along the walls of ravines when the water is low, curious round cauldron-shaped cavities with smooth sides may often be observed. These, known as **pot-holes** (Fig. 61), are formed by the grinding action of loose stones, which, caught in eddies of the water, are kept in rapid rotation. Of course the stones, in excavating the holes, are themselves reduced to sand or fine gravel, but other stones are swept in to supply their place. On the sides

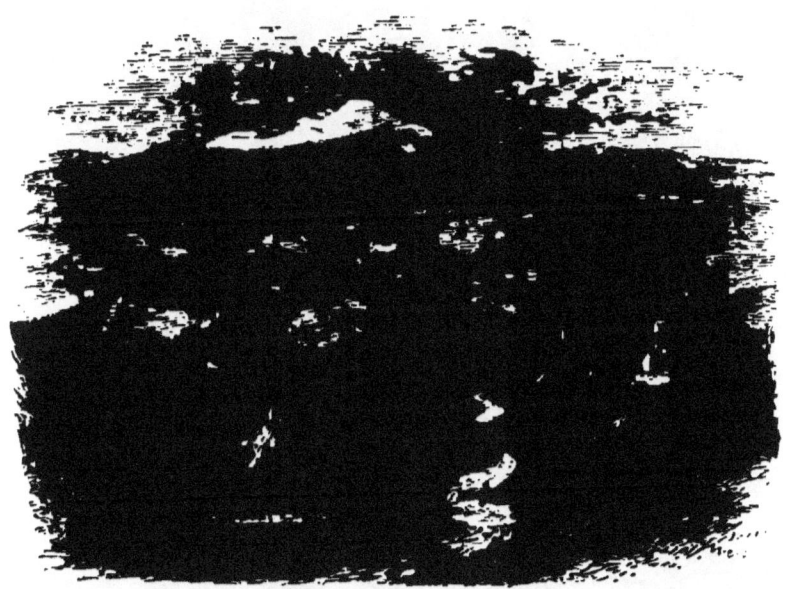

FIG. 60.—View of ravines cut by streams out of a table-land.

of many a narrow gorge through which a stream forces its way, traces of old pot-holes may be seen high above the present water-line. These mark former levels of the stream, and show how in course of time the rocky bed has been gradually dug out.

11. Another way in which a stream erodes its channel is by means of **waterfalls** (Fig. 62). It is not always possible to tell from what cause a particular waterfall may have been formed at first. Some original cliff or steep bank, when the stream began to flow, or some

harder mass of rock, encountered in the erosion of the ravine, may have determined it. But in what way soever it may have begun, a waterfall gradually creeps up stream, carving out a ravine between its original

FIG. 61.—Cascade and pot-holes of a water-course.

point of commencement and the point which it has now reached. This excavation is done by the recoil of the water and spray, whereby the rocks behind the bottom of the fall are loosened and precipitated into the whirlpool or rapid below. So long as these crumble down more easily than those at the top of the fall, the cliff over which the water dashes will remain precipitous. Slice after slice being cut off its face, the precipice over which the water tumbles will shift its place farther and farther up the stream, carrying with it the waterfall which, though slowly moving up the gorge, may appear stationary, because it will preserve from year to year the same general appearance. If, however, from any differ-

FIG. 62.—Section of a waterfall and ravine.

ence in the nature or position of the rocks, the rate of waste should become more rapid at the top of the cliff than at the bottom, the cliff, instead of overhanging, will then begin to retire at its upper part. The waterfall will now gradually grow less marked, until it will pass into the condition of **rapids**, that is, a shoot of water over a steep and rough part of the bed. Finally, these rapids may themselves be worn down, and all trace of the original fall will then disappear, except the gorge which it excavated during its recession.

12. Almost every large river, flowing through a hilly or mountainous region, illustrates these features of the erosive action of running water. Perhaps the most stupendous example is that of the Niagara River. The famous Falls of Niagara consist of two vast cascades separated by a small island, and having a united breadth of 950 yards and a height of 140 to 160 feet. It has been computed that 670,000 tons of water are poured every minute over these falls into the foaming torrent below, from which vast clouds of spray rise up into the air. Originally the falls stood at Queenstown, where the limestone forms a cliff above a great plain. Since that period, however, the cataract has slowly receded for about seven miles to its present position. That it is still moving upwards is shown by the large slices of rock that from time to time fall from the cliff over which the water rushes. The present rate of retreat has been computed at about one foot in the year. Probably this is an exaggerated estimate; if it be taken as an average for the past work of the river, somewhere about 35,000 years must have been required to excavate the ravine between the present Niagara Falls and Queenstown.

13. Every stream, then, which drives along sand and gravel on its bottom, is busy with the work of erosion. If even in ordinary weather this action may be perceived, how much more stupendous must it be in floods, when every little runnel is swollen, when earth, sand, and stones are swept by rain off the ground, and when the rivers, rising high above their ordinary level, and acquiring from the increase of their volume augmented velocity of flow, rush over the land and bear their vast burden of detritus down to the sea. A river may do far more of its erosive work in a few hours of flood than in many days or weeks of its ordinary flow. Hence this kind of river-action reaches its maximum when the river attains its greatest body of water and highest velocity.

14. **River-Transport.**—The loose materials acquired

in the process of erosion are removed and variously disposed of by the streams. As long as the water has velocity enough, it keeps the sediment moving, and conveys it sometimes to great distances. Any check to the velocity causes some of the sediment to fall to the bottom. In considering the nature and amount of work done by rivers in the transport of mineral materials from the land, we must bear in mind that these materials exist not only in visible form, such as gravel, sand, and mud, but invisibly dissolved in the water. As every spring is busily employed in bringing up to the surface mineral substances which the water has dissolved out of underground rocks, and as rain and streams are similarly engaged above ground, a vast quantity of dissolved material must be conveyed into the sea. It is not difficult to make an approximate estimate of the amount of invisible mineral substance thus carried by a river. The amount of water discharged by the river must be ascertained, likewise the average proportion of mineral ingredients contained in a gallon or other given quantity of the water. The one sum multiplied by the other will give the required result. The celebrated chemist, Bischoff, calculated that the Rhine carries past Emmerich every year enough carbonate of lime, chemically dissolved in its water, to form 332,539 millions of oysters of the usual size. If all these oysters could be put together they would form a cube measuring 560 feet in the side. The river Rhone is estimated to carry past Avignon every year 8,290,464 tons of dissolved salts in its water. The annual discharge of the Thames past Kingston, which stands a few miles above London, is estimated at 548,230 tons of mineral matter, two-thirds of which is carbonate of lime. It has been computed that the rivers of England and Wales carry every year into the sea 8,370,630 tons of solids in solution. If this quantity were entirely dissolved away from the surface, it would be equivalent to a general lowering of the surface of the country at the rate of one foot in 12,978 years.

15. But by far the largest amount of mineral matter borne by rivers from the land is in the form of mechanical sediment—gravel, sand, and mud. Every river is more or less muddy. After heavy rain even the clearest brook is discoloured by the earth it carries down. Mere discoloration, therefore, is a proof of the constant transport of sediment by running water. The amount of material thus transported depends partly, of course, upon the carrying power of the river, which is regulated by its volume and velocity; partly upon the nature of the soil and rocks of its drainage-basin, whether they happen to be earthy and easily worn away, or the reverse; partly upon the distribution of the rain-fall, whether it is spread over all the seasons of the year, or crowded into a few weeks or months, so as to produce a swollen and muddy torrent while it lasts; and partly, where the river takes its rise from a glacier, upon the quantity of mud which escapes from the melting end of the ice. (Lesson XXVIII.)

16. A stream having a current of about half a mile in the hour, which is a comparatively feeble flow, can carry along ordinary sandy soil suspended in the water. With a velocity of twelve inches in a second, which is about two-thirds of a mile in the hour, it can roll along fine gravel, while, when the rate rises to three feet in a second, or a little more than two miles in the hour, it can sweep away slippery angular stones as large as an egg. We can readily understand that in torrents, with a steep slope and high velocity (Lesson XXV. Art. 22), the power of transport must be enormous. Huge masses of rock, as large as a house, have been known to be moved during heavy floods.

17. It is evident, therefore, that in a rapid river or brook, the mud which discolours its water represents only a part of the sediment carried down. A great deal of sand and gravel, or even coarse shingle, is at the same time being pushed along the bottom. This material cannot, indeed, be seen, but the large stones may sometimes be heard rattling against each other, as they are rolled onward by the current.

18. Measurements and estimates have been made of the proportion of sediment in the water of different rivers. This proportion varies, of course, in different seasons, being greatest during floods, and least when the rivers are low. The Ganges, during its four months of flood, is stated to contain one part of sediment in every 428 parts by weight of water, while the mean average for the year is one part in 510. In the water of the Irrawaddy the proportion was found to be one part in 1700 by weight of water during floods, and one part in 5725 during the dry season. In the water of the Mississippi the average proportion was determined to be one part in 1500 by weight, or one part in every 2900 by volume of water. The Danube has been found to contain a mean proportion of one part in 3060 by weight of suspended matter, and during extraordinary floods discharges into the Black Sea as much as 2,500,000 tons of silt in twenty-four hours.

19. In any adequate estimate of the total discharge, the coarse heavy sediment which is pushed along the bottom must also be allowed for. In the case of the Mississippi, this moving layer has been estimated to deliver into the Gulf of Mexico the vast amount of 750,000,000 cubic feet of earth, sand, and gravel every year.

20. Having ascertained the average quantity of mineral matter suspended in the water, or pushed along the bottom, and having estimated the average amount of water carried by a river into the sea, we may easily obtain, by multiplication, the total quantity of sediment removed from the land by that river in a year. Thus the Rhone is estimated to carry into the Mediterranean

every year rather more than 600,000,000 of cubic feet of sediment. The discharge of the Danube into the Black Sea has been determined to be 67,760,000 tons of silt annually. The mean yearly amount of solid matter carried in suspension by the Mississippi into the Gulf of Mexico is estimated to be about 362,723,000 tons; and this, including the coarse sand and gravel which are pushed along the bottom, would make a column one mile square and 268 feet high. We may form some notion of this amount of material by supposing that 1000 merchantmen, each laden with 1000 tons of it, were to arrive every day for a whole year at the mouth of the Mississippi and discharge their cargoes into the sea. They would little more than equal as carriers the work of this single river.

21. But many rivers greatly exceed the Mississippi in the proportion of solid matter which they transport. During the rainy season in India the streams become torrents of mud. Dr. Livingstone in his African travels came upon "sand-rivers"—currents of sand, moving along with a comparatively small amount of water. In trying to ford them, he felt thousands of particles of sand and pebbles of gravel striking against his legs, even in dry weather, and he saw that after the rains the quantity of detritus removed by these streams must be enormous.

22. The amount of sediment carried down by a river to the sea in a year represents the extent of loss which the region drained by the river has sustained within that time. Knowing the quantity of sediment and the area of country from which it has been derived, we can determine the amount by which the general surface of the river-basin has been lowered. Thus at its present rate of work, the Mississippi reduces the general level of its drainage area $\frac{1}{6000}$ of a foot annually, or one foot in 6000 years. Could this rate of denudation be continually kept up over the surface of the land, which is computed to rise to an average level of about 1000 feet

above the sea, a whole continent might be reduced to the sea level in about 6,000,000 years. Such calculations are of importance in showing that the present surface forms of the land must be continually changing, and cannot, therefore, be of comparatively high antiquity.

23. **River-Deposit.**—All rivers, then, are constantly busy grinding down and transporting gravel, sand, and mud over the surface of the land. To ascertain what becomes of all this material, let us in imagination again follow the course of a river, from the mountains to the sea, and watch how the sediment is disposed of by the way.

24. When running water has its velocity checked, it loses some of its power to transport sediment, which then partly sinks to the bottom. This may happen when a stream enters upon a gentle slope or plain, where it must move more slowly, or when it joins a larger and more gently-flowing current, or when it falls into still water, like that of a lake, or into the sea. Hence, during its course, as well as at its termination, a river necessarily encounters many obstacles to its progress, by which it is compelled to slacken its pace and to drop some of its sediment. The general name of **alluvium** or "alluvial deposits" is given to accumulations of detritus by running water.

25. Beginning among the mountains, we meet with abundant examples of this arrest of the detritus which the torrents have been sweeping down the declivities. A steep slope is often deeply trenched with gullies that have been torn out of the soil or rock by torrents. Where, at the bottom of these ravines, a strip of more level ground stretches in front of them, they each heap up a pile of rubbish, which the headlong brooks, checked in their flow by the change in their angle of descent, have been forced to throw down. Among mountains, so numerous may be the torrents, and so vast the heaps of gravel and stones which they tear out of the hill-sides and strew over the

slopes and valleys, that it is in many places difficult to maintain roadways, for these are liable at one point to be buried under huge masses of debris, and at another to be swept away by a flooded stream.

26. But only the coarser kind of sediment is usually arrested at the foot of such mountain slopes. The finer parts, and even some of the rougher gravel, are carried farther into the valley, where the various brooks unite into one main stream. At every point where this stream is checked, an accompanying deposit of sediment occurs. Among its many serpentine windings the current, though rushing briskly round the convex side of a bend, is thrown into an eddy or slack water on the concave side, and there a bank of sand or shingle will generally be found.

27. When the stream is flooded, it not only fills its ordinary channel, but rises and overflows the flat meadows on either side. These tracts of level ground, by diminishing the velocity of the overflowing water, compel the stream to drop some of its sand and mud upon them. Should they be covered with herbage or brushwood, the leaves and branches act as filters, and clear the water by retaining the sediment. So that when the stream has subsided, the inundated ground is found to have received a coating of fine silt, or even, it may be, of coarse gravel. The flat land which lies on either side of a river, and is liable to inundation when the water rises, is termed the **flood-plain.**

28. If, then, each flood adds to the height of the flood-plain, the time will come when, even at the highest flood, the water will not be able to overspread the plain. This result arises not only from the heightening of the plain by deposit of sediment, but also from the gradual deepening of the stream-channel by the scour of the current. As the channel is deepened, the current continues to eat into its banks, curving from side to side, and forming a new plain at a lower level. This process has been in progress for a long time in most river-valleys. A succession of terraces marking former levels of the river-floods may be seen rising even up to heights of several hundred feet above the present rivers. The accompanying figure (Fig. 63), representing a section drawn through such a river valley (S. S.), shows the relation which the low level terrace or present flood-plain (3) bears to the higher, and, of course, older ones (2, 1). In some of the latter, remains of primitive man, such as chipped stone spear-heads and other implements, have been met with in different parts of Europe, showing that when the rivers flowed at the level of these higher terraces, a rude human population already existed in countries wherein the river-valleys have undergone hardly any appreciable change within historic times.

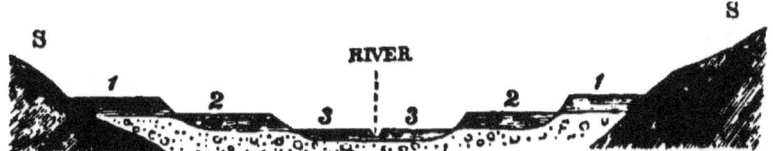

FIG. 63.—Terraces of gravel, sand, and mud, left by a river.

29. While still in the mountainous or hilly part of its course, a river may have to traverse one or more lakes. Each of these sheets of still water, by arresting the current, compels it to drop its burden of sediment. Lakes filter river-water, which, leaving on their floors its sand and mud, issues at the lower end quite clear (Lesson XXVI.). The chief deposit takes place where the river enters the lake. By degrees that portion of the lake is filled up and converted into a plain, which, after being gradually heightened by the river-floods, at last comes to lie above the flood-limit. In this way the upper end of the Lake of Geneva has been so diminished by the deposits of the Rhone, that a Roman port, still called Port Valais, is now nearly two miles from the edge of the lake, the intervening space consisting of meadows and marshes.

30. Lakes must evidently in course of time be filled up by the earth and sand washed into them. This has already been the fate of many. The once united lakes of Thun and Brienz in Switzerland have been separated by a tract of land, formed by the deposits brought down from either side by the streams at Interlaken. In Great Britain, as well as generally throughout Northern Europe, every stage in the disappearance of lakes is abundantly shown, from the mere tongue of sand encroaching upon the edge of a deep mountain tarn up to the flat moss or fertile meadow, which marks where a former lake has been silted up.

31. When a river enters the lowlands, the diminished slope of its course causes it to flow with a weaker current, and therefore to drop some of its burden of mud and sand as it moves along. Winding to and fro, it cuts down its banks at one part, and heaps up sediment at another, so that in course of time, the whole of a wide plain comes, piece by piece, to be levelled by the shifting stream. The plain is formed, indeed, out of sediment carried down by the water from higher ground. A well or pit, sunk anywhere over its surface, shows that beneath the soil there lie layers of water-worn silt, sand, or gravel, like the materials now being transported and deposited by the river.

32. When its plain is long and the seaward inclination slight, the flow of a river may be so lazy that, instead of scouring out its channel, it may at last be unable to prevent the sediment from sinking to the bottom and actually heightening the bed of the stream. During floods, the chief deposit of silt takes place on the banks. These are consequently raised in level more than the plain beyond, which, not being so liable to inundation, receives less sediment. In this manner, the river gradually comes to flow at a higher level, between broad embankments of its own building. From time to time it breaks through the lower or weaker parts of these

embankments, and inundates the plain, perhaps even scouring out here and there a new bed. In cultivated regions, like those watered by the courses of the Po and Adige, in the plain of Lombardy, and by the lower portion of the Mississippi, much care is needed to strengthen and heighten the banks, with the view of averting inundations. Some rivers have so heightened their channels that their surface flows during floods at a higher level than the streets of towns on their banks.

33. The most famous example of such a broad plain, formed by the accumulation of sediment brought from the interior of the land and laid down by the long-continued flow and overflow of a river, is Lower Egypt, which even in ancient times was recognised to be a "gift of the Nile." At midsummer every year, this river begins to rise and to cover the flat land on either side of its course. The inundation is at its height in about three months, and then the waters, after remaining stationary for nearly a fortnight, retire to their former level. This annual rise corresponds to the rainy season in the mountainous tableland of Abyssinia (Lesson XXV. Art. 18). When the monsoon blows from the Indian Ocean it brings torrents of rain, which rush down the rugged slopes of that country and sweep along a prodigious amount of mud. Thousands of swollen and muddy streams are at last united in the Blue Nile, which carries this vast body of discoloured water down into Egypt. After the inundation has ceased, the overflowed ground is found to be covered with a thin coating of rich fertile mud. It has been estimated that the thickness of this annual deposit does not exceed that of a thin sheet of pasteboard, so that a depth of only two or three feet of the soil of Lower Egypt represents the continuous deposits of a thousand years. The increase of the Egyptian plain evidently takes place at the expense of the high grounds of Abyssinia. It is the finer particles, worn away from the rocks of these uplands, which form the mud spread over Egypt. Here we see how the erosion,

transport, and deposit of materials by running water combine to build up a plain and renew its fertility.

34. The great plains of India furnish, likewise, admirable illustrations of the way in which the debris of the mountains is spread out on low grounds by rivers. The valleys of the Indus, Ganges, and Brahmaputra have been filled up by the accumulations carried down from the Himalaya chain by these great rivers. The Tigris and Euphrates have combined to fill up the upper half of the valley of which the Persian Gulf is the still remaining lower half. On the American continent, also, this process is exhibited on the most stupendous scale. Much of the eastern borders of the United States is a plain formed by the deposit of sediment washed off the land. Many thousands of square miles of nearly level land in the valleys of the Mississippi and its tributaries are underlaid by alluvial deposits. The valley of the Amazon, with its vast forests or *silvas*, forms so long and so level a plain that ships can sail up the river to the very foot of the Andes, a distance of 2000 miles inland from the sea.

35. Lastly, it remains to inquire how the sediment borne down by rivers is disposed of when it reaches the sea. Many rivers have, at their mouths, what is called a **bar**, that is, a ridge of gravel or sand stretching across the channel, and always under water. From what has been said already in this Lesson, the origin of this bar will now be understood. It is evidently due mainly to arrest of the sediment, where the river current is checked by coming in contact with the sea. The coarser gravel and sand, pushed along the bottom of the river channel, now encounter the opposition of the salt-water, over which the lighter fresh water of the river flows onward. The sea piles up fresh materials upon the bar from the outside, while, on the other hand, the river when flooded drives its bar farther seaward. Hence this barrier to navigation at the mouth of a river is continually shifting its position

and altering its shape and size, according as the action of the river or that of the sea predominates.

36. Some of the larger rivers of the globe exhibit at their mouths on a vast scale the same operations that take place where streams enter lakes. The sediment deposited by them in the sea has gradually filled up bays or gulfs and converted them into flat land, or has pushed its way out to sea. This advancing tract of river-formed ground is usually triangular in form, the apex pointing up the river. It was this resemblance to the Greek letter Δ that suggested the name of **delta**, as applied to these accumulations at the mouths of rivers. Down to the head of its delta a river (Fig. 52) is usually augmented by tributaries from either side, and does not branch, except for a limited space, as where it encircles an island in its course. Its current thus becomes more and more ample. But when it reaches the delta it begins to subdivide, and continues to branch out, until in some cases the flat plains and marshes are traversed by innumerable tortuous channels of water. (Lesson XXV. Art. 6.) By this ramification over the flat ground the velocity of the current is diminished and the deposit of sediment is facilitated, consequently the channels are being continually filled up, while new ones are cut through the soft alluvial earth and silt. Two or more main branches of the river carry out the chief part of the water and sediment to sea, and at their mouths the increase of the delta is most apparent.

37. The general form of a typical delta is shown by that of the Nile (Fig. 52) which enters the tideless Mediterranean. The Mississippi, on the other hand (Fig. 64), brings down such an amount of sediment that not only has it filled up its valley, but it now pushes tongues of alluvial land far out into the Gulf of Mexico. The average rate of advance of this delta has been estimated to amount to 86 yards in the year. The Tiber throws forward its delta at the annual rate of about 12

to 13 feet. The delta of the Po has increased at such a rate that the port of Adria, which stood on it and was so important in Greek and Roman times as to give its name to the Adriatic sea, is now 14 miles inland.

38. The deltas of some of the larger rivers of the globe are of enormous size. That of the Mississippi embraces an area of about 40,000 square miles. That

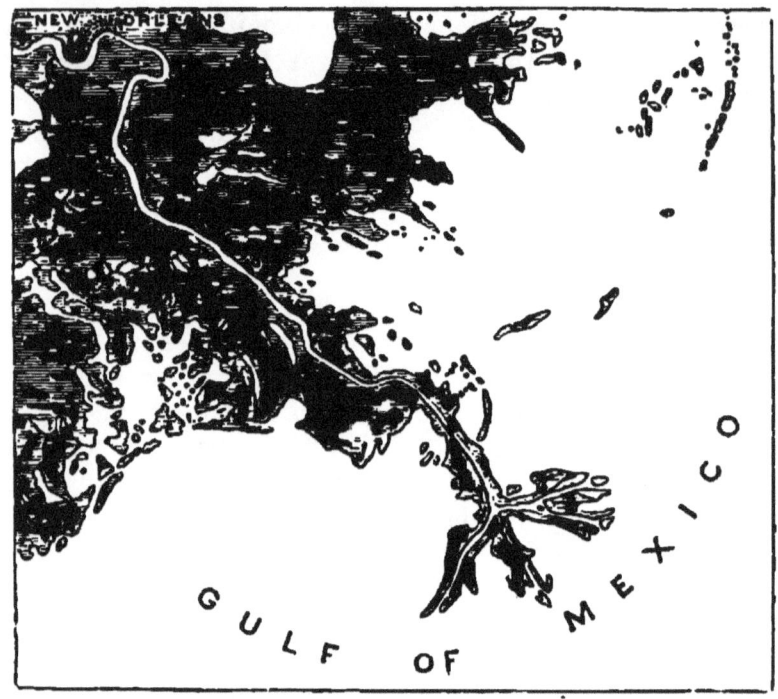

FIG. 64.—Delta of the Mississippi.

of the Ganges and Brahmaputra is as large as the whole of England and Wales. Their vast superficial extent indicates the high antiquity of these deltas. But we must remember also that, as they have been formed by the gradual filling up of gulfs of the sea, a prolonged period of time, where these gulfs were deep, would be occupied in bringing up the level of the bottom to that of the surface, before any advance of the plain of the delta could take place. Where the tides and currents of the sea interfered in such a way as to sweep away much of the sediment as it arrived, the time required would be still further increased. The delta of the Ganges has been bored into at Calcutta, and its deposits, consisting of sand, gravel, silt, and layers of vegetation, were found to be more than 400 feet thick.

39. In cases where the form of the coast is unfavourable to the formation of deposits, or where the rivers flow with sufficient velocity to carry out their mud to sea, or where marine currents sweep past the river-mouths and carry away the sediment, deltas do not occur. The Amazon and La Plata, for example, form no deltas. But they nevertheless pour a vast quantity of fine silt into the Atlantic. Even for a distance of 300 miles from the mouth of the Amazon the sea is said to be perceptibly discoloured by the mud of that river.

LESSON XXVIII.—*The Waters of the Land—Part V.*

Frost, Snow-fields, Glaciers.

1. The moisture of the air, when the temperature sinks to the freezing-point, passes into ice (Lesson X. Art. 36). We have now to follow the course of this ice when it falls from the air to the land. We must further consider how the waters of the land are affected when cooled down below the freezing-point; for, changed into solid ice, they cannot but acquire new and different powers from those which have been the subject of the last four Lessons. What part, then, does the frozen water of the land play in the general plan of the globe? Some answers to this question may be found in the study of the behaviour of three forms of ice on the land—Frost, Snow-fields, and Glaciers.

2. Frost.—Most substances suffer contraction from cold, and consequently increase in density. A cubic foot of pure water, for example, at a temperature of 40° Fahr. weighs more than the same quantity at 60°: or, in other words, a vessel exactly filled with water at 60° would be found to be not quite full were the water cooled down to 40°.

3. But a remarkable change takes place during the further cooling of the water. When the temperature reaches 39·1° Fahr., contraction ceases. This is, therefore, the temperature at which pure water is heaviest, and is called the *point of maximum density* of fresh-water. Below this point, the water expands, and instead of sinking, remains on the surface where it is at last converted into ice when the temperature has fallen to 32°. Hence it is that sheets of water are not frozen to the bottom. The ice being lighter than water, floats on the surface. As the cold continues, the first thin crust of ice is thickened by additions to its under side, until after a severe winter it may be a foot or two in thickness. In this way, during a long and hard frost, rivers and lakes become so firmly frozen over that carts and heavy waggons may be drawn across them, though the water with its living inhabitants remains still liquid underneath. Canals and rivers which, for most part of the year, are highways for boats and barges, become, in such cold winters, thoroughfares for carriages and sledges, as well as passengers on skates or on foot.

4. The parts of the earth's surface where this curious transformation may be ordinarily seen in winter are indicated on Plates IV. and V. by the position of the isotherm of 32°. In the northern hemisphere, over all the ground lying to the north of that isotherm, and in the southern hemisphere, on all the ground lying to the south of it, the waters are frozen during winter. But even in much more temperate latitudes, the cold is occasionally severe enough to make the rivers and canals passable on foot, and what

is more remarkable, actually to encrust the sea with ice. Thus in the year 401, and again in 642, the winter was so intensely cold in Southern Europe that the Black Sea was entirely frozen over; in 850 the Adriatic was covered with ice; while in 1233 and in 1314 the rivers in northern Italy were likewise frozen. In the year 1205 a severe frost, lasting for three months, bound up the soil of England so that it could not be ploughed, and froze the canals, rivers, and ponds. The last time that the Thames at London was passable on ice was in the year 1814. Thousands of passengers crossed the river on foot between London and Blackfriars Bridges, and numerous wooden booths and tents were erected on the ice, together with merry-go-rounds and other amusements, so that the busy concourse of people was called "Frost Fair."

5. The freezing of sheets of water on the land may take place so equably, and the ice may disappear again so quietly, that after the frost has passed away, no trace may be left of its having occurred at all. But should a frozen lake break up under a storm of wind, large masses of ice will be driven ashore, and may push up sand, gravel, and stones lying at the water's edge. Such heaps of ice often take some weeks to thaw, and after they have melted, the stones and sand which they had driven ashore are found scattered over the ground, as may be seen on a great scale on the shores of the large Canadian lakes. In the lakes and rivers of Canada also, another result of severe frosts may be observed. Blocks of stone lying in shallow water are frozen into the ice, and when the spring thaws set in are actually lifted from their places by the fringes of ice which have formed round them, and are floated away to some other part of the shore, or may even be driven out into the deeper water, so as at last to be dropped there. This transport of stones has been particularly observed on the shores of the St. Lawrence.

6. An indirect result of the freezing of rivers, but one

of great importance, is the accumulation of the ice in bars, whereby disastrous floods are produced. When the ice breaks up in a rapid thaw, huge blocks of it, borne down by the current, and piled upon each other, may be driven together in such a way as to dam up the river. The water accumulates behind the ice-barrier, and at last bursts through it, sweeping with prodigious force down the valley, and carrying destruction far and wide in its course.

7. The effects of frost, however, are to be traced not merely on sheets of water on the land. Rain, falling upon the land, soaks through soil and rocks, which consequently contain abundant water in their pores and cavities. When severe frost sets in, this water freezes. The soil becomes hard, so that even muddy places, where one would have sunk deep in mire, can now be safely and easily traversed on foot. A jug of water exposed to severe frost is apt to be split. In insufficiently warmed dwelling houses, water-pipes in like manner burst during frost. The reason of these mishaps is to be sought in the expansion of water in the act of freezing. When passing from the liquid to the solid state, water undergoes a sudden and remarkable expansion, amounting to about one-tenth of its volume. At the moment of change it exerts great force upon the sides of the vessel or cavity containing it, and if it cannot readily escape it will do its best to burst these sides, that it may find room for its increase in bulk. Now what takes place in a water-jug or pipe, goes on also among the little particles of water enclosed between the grains of soil and rocks. Frost by expanding the water, pushes these grains aside. On a winter morning, after a night of sharp frost, we may notice that the expansion has been great enough to force the little pebbles of a roadway out of their places. So also, in countries like Canada, where the winters are extremely cold, wooden fences are, in the course of a year or two, twisted out of the ground.

8. It is not until a time of thaw that we are made fully aware of what the frost has done. So long as the cold continues, the separated particles of soil are kept together by the ice, which binds them into a hard solid mass. But when this ice melts, the grains of sand and earth become loosened from each other. Walking on a road or over a ploughed field after frost, we find that this loosening has been carried so far that the ground has become coated with mud. Indeed the millions of little ice crystals, which frost wedges in among the grains of soil, have much the same effect as if the earth were ground down in a grinding-mill or mortar. They pulverise it, that is, reduce it to powder, and thus lay it more open to the branching roots of plants, which obtain so much of their sustenance from the soil. Farmers are in the habit of ploughing their land before cold weather sets in, so as to leave the upturned loam exposed to the beneficial effects of succeeding frosts.

9. The powerful mechanical effects of frost are so well seen in soil, because of the abundant moisture retained among the particles of which soil is composed. But any porous rock which contains sufficient water, and is exposed to a great enough cold, may show the same kind of disintegration. Hence in countries where the winters are severe, ordinary building-stones and mortar are found to peel off in successive crusts, or to crumble down into powder, after frost has given way to milder weather. Even in the comparatively mild winters of Britain, this may be constantly seen; in the severer climate of North America it is a serious and costly evil, since it prevents the use of many kinds of building-stone, which in the absence of frost would be very valuable.

10. Passing now from the water retained and frozen among the pores of soil and of rocks, let us consider further the result of the formation of ice in the larger cracks and crevices of cliffs and crags. To realise how abundant these natural lines of division are in all rocks,

look at any exposed face of rock such as a sea-cliff, the ravines of a river, or the precipitous sides of a mountain. No matter what may be the kind of rock, it is traversed by many parallel and intersecting divisional planes or "joints," which serve as channels, wherein the surface water descends underground and re-ascends to form springs (Lesson XXIV. Art. 17). When frost becomes so severe as to penetrate beyond the mere outer crust of rock, the water contained in the external parts of the joints freezes. This must often take place in cavities where there is little room for expansion, and where, therefore, the ice exerts its great force in pushing the walls asunder. Winter after winter the process is repeated, until at last the portion of rock on the outside gets so far wedged off from the rest, that it loses its balance and falls to the bottom of the cliff. In all countries subject to intense frosts, the bases of cliffs and crags may be seen to be strewn with large rough blocks, which have in this manner been loosened from their places above. In the valleys among mountains that rise high up above the snow-line, the operations of frost are powerful in splintering the crags and pinnacles, and giving them the sharp spiry forms they so often assume. (Lesson XXIX.)

11. **Snow-fields.**—Wherever the land rises above the snow-line, it is buried under a permanent sheet of snow, from which only the higher and steeper mountain peaks project. In some regions, as on the table-land of Norway, a broad and tolerably level plateau allows the snowy covering to spread in a vast undulating sheet, three or four thousand feet above the level of the valleys. Standing upon one of the heights at the edge of such an expanse of snow, we see what looks like a white frozen plain, with no limit save the line where it meets the sky. In other parts of the globe, where rugged groups of mountains tower above the snow-line, and there are consequently no such level tracts, the snow accumulates in the hollows

and on the higher slopes. The semicircular ranges of mountain cliffs and crests often enclose vast basin-shaped depressions, each of which becomes a gathering place for the snow. To all the permanent sheets of snow, whether occurring on table-lands or in the hollows of mountains, the general term of **snow-fields** is given.

12. In these regions, since the moisture falls from the air as snow rather than rain, and the heat of summer is insufficient to melt it all, the quantity of snow would increase indefinitely were there no provision for the removal of the surplus. The thickness of snow on some snow-fields must amount to many hundred feet. Greenland, for example, is almost wholly buried under one vast snow-field, so deep as to cover over the inequalities of the surface of the land almost as completely as the furrows of a ploughed field are lost beneath a heavy snow-wreath. The Antarctic regions are believed to be buried under a mantle of snow passing down into solid ice, having a thickness of 10,000 feet or more.

13. There are two ways, besides melting and evaporation, in which the snow-fields get rid of their excess of snow; these are **avalanches** and **glaciers**. Where the edges of a snow-field overhang steep slopes, portions of the more or less consolidated snow from time to time break off, and rush with a noise like thunder and a terrific force into the valleys, tearing up the soil, sweeping down loose blocks of stone, uprooting or breaking trees, and carrying destruction as far as they reach. These snow-falls are called avalanches. In Alpine countries, forests which lie in the pathway of such descending masses of snow are carefully preserved as barriers to protect the meadows and villages from ruin. Roads which pass along the base of snowy mountains require in some places to be covered over with a strong archway of masonry to protect them from the disasters caused by frequent snow-falls, which would not only sweep away every traveller and carriage in their course,

but would even tear up the road itself, or bury it under heaps of earth and stones.

14. **Glaciers**, however, are the chief means whereby the superabundant snow above the snow-line is removed. While the snow forming the surface of a snow-field is loose and open in texture, like that which covers the ground in winter and disappears in spring, the under portions become more and more close and firm under the pressure of the overlying mass. This compacted snow (*nevé* or *firn*, as it is called in Switzerland) passes by degrees into clear blue ice, as the imprisoned air is more and more completely squeezed out of it. If the snow-fields lay on perfectly level table-lands, there would be no general movement of the snow except along the edges of the plateaux, where the gathering snow-sheet would break or slide off into the valleys. But since the surface of the land has a more or less marked inclination from its axis or water-shed, the snow, by virtue of the action of gravity, must slide downward even upon a very gentle slope. It is chiefly during this movement that the air is pressed out, whereby the loose, white, opaque snow is converted into solid blue, transparent ice. Having acquired a slow, sliding motion, the mass of snow seeks the lowest levels. It therefore moves downward into the heads of the valleys that ascend into the snow-fields. Each of these hollows becomes a kind of reservoir in which the snow, pressing downwards from each side as well as from behind, accumulates to a great depth, and is so jammed up between the mountain slopes as to take the form of solid ice from bottom almost to top. Driven onwards by the pressure of the advancing mass behind and by its own gravity, this ice fills up each valley sometimes to a depth of several hundred feet, and for a distance of many miles. Such a tongue of ice, proceeding from a snow-field above and descending below the snow-line, is termed a **glacier**. In the accompanying drawing (Fig. 65) two glaciers are

seen descending from one of the great snow-fields of Arctic Norway. In one case, the ice almost reaches the sea; in the other the glacier is smaller, and does not proceed so far from its parent snow.

15. A glacier, then, represents the escaping drainage of the snow-fall, above the limit of perpetual snow, as a river does the surplus drainage of the rainfall. Its size must depend upon that of its enclosing valley, and on the

FIG. 65.—Snow-field and glaciers of Holands Fjord, Arctic Norway.

extent and declivity of the gathering ground of snow from which it issues. In the Alps, for example, the Great Aletsch Glacier extends about fifteen miles down its valley. Sometimes a little glacier fills up a high recess on the flank of a mountain, and never reaches even to the nearest valley. In other cases the glacier descends far below the snow-line, even into the region of meadows, corn-fields, and gardens. The lower glacier of Grindelwald, in the Bernese Oberland, reaches to a point about 3500 feet below the snow-line on the north side of the Alps.

16. To realise as clearly as possible the general

appearance presented by a glacier, let us suppose ourselves placed at the lower end of one of those among the Alps. On either side of the valley the slopes are clothed with pine. Patches of green pasture catch the sunlight in the hollows and on the lower projecting hills; while around us lie scattered cottages and bright meadows. In front stands the abrupt end of the glacier—a

Fig. 66.—View of a glacier, with its lines of rubbish (moraines) and the river which escapes from its end. Ice-worn hummock of rock and transported stones are shown in the foreground.

steep, craggy face of ice, from the base of which issues a river of pale muddy water. Numerous large blocks of rock are scattered about on the valley-bottom below the ice. The ground there, indeed, is mainly composed of coarse shingle, like that which forms the bed of the present river. Even on the ice itself we may see heaps of stones, some, perhaps, poised just on the verge of the last steep slope of the glacier, whence they must soon roll down to join the crowd of others that have preceded them. Looking into one of the deeper rents in the ice, we see it to be of wonderful purity, and of the most exquisite blue colour. And yet most of its outer surface may be so obscured by earth and stones, that at first, perhaps, we can hardly be persuaded that it is ice at all.

17. We ascend to the surface of the glacier, either by mounting among the broken cliffs of ice forming its abrupt front, or by choosing a safer and easier path up the slope on one side of the valley. The ice is now seen to lie as a great sheet, filling the bottom of the valley from side to side, and stretching far up into the heart of the mountains. Its surface has at first a gentle slope, and is comparatively smooth, though many glaciers even at the lower end present a marvellously rugged aspect, like that of a tempestuous sea suddenly frozen. Ascending the valley, we notice that the surface accumulations of earth, gravel, and stones are especially abundant along the sides of the glacier, and likewise in one or more ridges along its centre. During the day, when the sun shines out warmly, the surface of the ice is thawed, and consequently abundant little runnels of water flow over it. At night, when these are frozen, the glacier becomes once more silent.

18. In several respects a glacier resembles a river; and this resemblance is closer than might at first appear. If the position of any prominent object on the surface of the ice be observed with relation to some fixed point on the bank, or if, as in the original survey of the Mer-de-Glace by the late Principal Forbes, a transverse series of stakes be driven into the ice across the breadth of the glacier, and their position be observed at intervals from the bank, the ice is found to be slowly moving down the valley. This motion is faster in the centre than at the sides, because the friction of the sides impedes the flow of the ice. The average rate of movement on the Mer-

de-Glace was thus ascertained to be, during summer and autumn, from 20 to 27 inches in the twenty-four hours at the centre, and from 13 to $19\frac{1}{2}$ inches at the side.

19. Bending from side to side, and travelling over an uneven, rocky floor, the glacier is split across by rents called *crevasses*, which often open and form wide and deep chasms, extending sometimes down to the very bottom of the glacier. Stones and earth lying on the surface of the ice frequently fall into these fissures, and thus reach the floor of the valley beneath the glacier, or are imprisoned in the ice, when the sides of the yawning crevasses are pressed together again as the glacier moves onward.

20. When a river reaches a steep and rocky part of its bed it forms a rapid; when it comes to a cliff, it takes the shape of a waterfall. Ice, not having the same mobility as running water, cannot so easily adapt itself to the irregularities of its channel. But it shows these irregularities in a very marked way. In the course of the glacier which we have supposed ourselves to be visiting, after, perhaps, some miles of slow ascent, we come to a precipitous slope, where the ice, completely shattered by innumerable fissures, rises into pinnacles and sharp crests of many fantastic forms. Could we watch that tumultuous descent of broken ice for a long enough time, we should find it to be all in slow motion downwards. It is an icefall, and answers in the mechanism of a glacier to the waterfall in that of a river. Underneath it, the bottom of the valley is steep or precipitous, and the ice, unable to descend the declivity in one unbroken sheet, is cracked and splintered in this wonderful way. But just as a river, no matter how much it may have been tossed into foam by its descent in a waterfall, speedily takes its usual shape and rolls onward as if no sudden plunge had so recently disturbed its current, so a glacier, though it may have been reduced, as it were, to fragments at one of these icefalls, soon reunites at the bottom, and again pursues its course as a solid and continuous sheet of ice. If the glacier is a large one it may

FIG. 67.—Plan of the Mer-de-Glace of Chamouni and its tributary glaciers, showing the way in which lateral moraines became medial.

receive tributary glaciers from valleys on either side. The Mer-de-Glace of Chamouni, for example, is formed by the united mass of several glaciers, as shown in Fig. 67. Each of these, as well as the main stream, may be traced upward until it is found insensibly merging into the snows which fill up the hollows in the higher part of the mountains.

21. One cannot trace the course of a glacier, and realise on the ground the size and character of those vast tongues of ice which carry off the drainage of the snow-fields, without wishing to know what task is allotted to them in the general economy of nature. The rivers which bear away the surplus water of the land are busy in a stupendous work,—wearing down the mountains and valleys, and strewing their debris over the plains, or sweeping it out to sea (Lesson XXVII.). Glaciers, too, are engaged in a similar task. They transport the waste of the mountains down to lower levels, and they erode the sides and bottoms of the valleys in which they move.

22. *Transport.*—Whence come the earth and stones which so darken and obscure the surface of the ice, and which, far below the snow-line, where the glacier melts almost at the edge of the meadows and gardens, lie piled in heaps on every side? From a good point of observation above the glacier we may notice that the heaps are not scattered wholly at random across the surface of the ice; but that there are long lines of stones that keep apart from but parallel to each other, and run along the length of the glacier. They wind to and fro, with the varying curves of the glacier in its course, until they are lost in the distance. In the drawing of a glacier in Fig. 66, for example, some of these lines of stones are seen both in the centre and towards either side. Following the central line on a glacier (sometimes there are several lines down the middle), we find at last that, at some higher part of the valley, it brings us to a point where two branches of the glacier join, and where the line of stones either continues up the side of one of the

branches, or divides into two, one portion keeping to the right-hand side of one of the tributary glaciers, the other remaining on the left-hand side of the other branch (Fig. 67). In every case it will be observed, if the glacier is followed far enough up its valley, that a line of stones in the middle of the ice comes really from the side, and is due to the confluence of two branches of the glacier. Owing to the irregularities in the slope and breadth of the glacier-channel, and the manner in which the ice is consequently driven to adapt itself to these, as well as owing to the melting of the surface of the glacier (Art. 25), the lines of stones are apt to lose their distinctness as they advance down the valley, until, after some great icefalls and abundant crevasses, the rubbish comes to be spread more or less over the whole of the glacier's surface, as it is towards the lower end of the Mer-de-Glace (Fig. 67).

23. These heaps of stones earth, and gravel lying on the ice are known by the name of **moraines**. When on the side of the glacier, they are termed *lateral* moraines; on the centre, they are called *medial;* at the end, where the glacier throws down its burden as the ice melts, they are spoken of as *terminal*.

24. In all cases, then, these moraine-heaps can be followed up the glacier to the base of some cliff or craggy mountain-slope, whence the blocks of rock have been derived. In such positions one sees how the fragments have been loosened by the severe and prolonged frosts of these high grounds, until, wedged off from the face of the cliffs, they have rolled down to find a resting-place upon the glacier below. Once on the ice, they are slowly borne down the valley, and dropped among the heaps of rubbish at the far end of the glacier. Much of the rubbish, however, falls down the numerous rents or crevasses in the ice, and reaches the bottom of the glacier, there to aid the ice in accomplishing another important part of its work (Art. 28).

25. Watching the progress of one of the large blocks as it travels with the ice, we discover what might not otherwise be so evident, that the surface of the glacier is continually being lowered by melting and evaporation. Take the case of a large flat stone, which, bounding from some high crag, has found a lodgment upon the ice. The portion of the ice lying below the stone is screened from loss by thawing and evaporation, but the surrounding parts of the glacier, not so protected, are insensibly wasted away. Consequently the stone begins, as it were, to rise out of the glacier. Its pedestal of ice

FIG. 68.—Glacier table—a pillar of ice supporting a block of stone.

continues to increase in height, but being exposed on the sides to sun and air, is lessened in diameter, until it becomes too slim to support the heavy burden of stone, which then tumbles down upon the surface of the glacier (Fig. 68). But as the general waste of the surface of the ice continues, the new position of the stone is soon marked by the rise of a new pillar of ice as before. The same block of rock may thus, in the course of its journey down the glacier, become the capital of several successive ice-columns. The same kind of testimony to the remarkable lowering of the surface of the glacier is shown by the long parallel moraine mounds. Looking at one of these ridges, and even climbing and standing on it, we might suppose it to consist of fragments of stone throughout. But by pulling down some of the loose blocks, we find that solid ice lies immediately below. It is in fact a ridge of ice with a coating of debris which has protected it from the general waste, as the detached stones screen the ice-pedestals beneath them.

26. In most valleys with glaciers in them, large blocks of rock may be observed above the present limits of the

Fig. 69.—The Pierre-à-Bot, near Neufchâtel (J. D. Forbes).

ice, but in places to which at one time the ice evidently reached. They occur poised sometimes in the most precarious positions, as if a man's strength would be sufficient to dislodge and send them down the slope. These **perched blocks**, as they are called, furnish good proof of the former extent of the glaciers. By their means, for instance, combined with other evidence, it can be proved that the glaciers of the Alps at one period filled up the Swiss valleys, and even spread over the broad plain of Switzerland, between the Bernese

Oberland and the Jura. On these vast ice-rivers blocks of granite from the Mont Blanc group of mountains were transported across what is now the valley of the Rhone and the Lake of Geneva, and were stranded high on the sides of the Jura range. The accompanying figure (Fig. 69) by Principal Forbes, representing one of these **travelled blocks, erratics**, or, as the Swiss call them, **foundlings**, shows the great size which some of them attain.

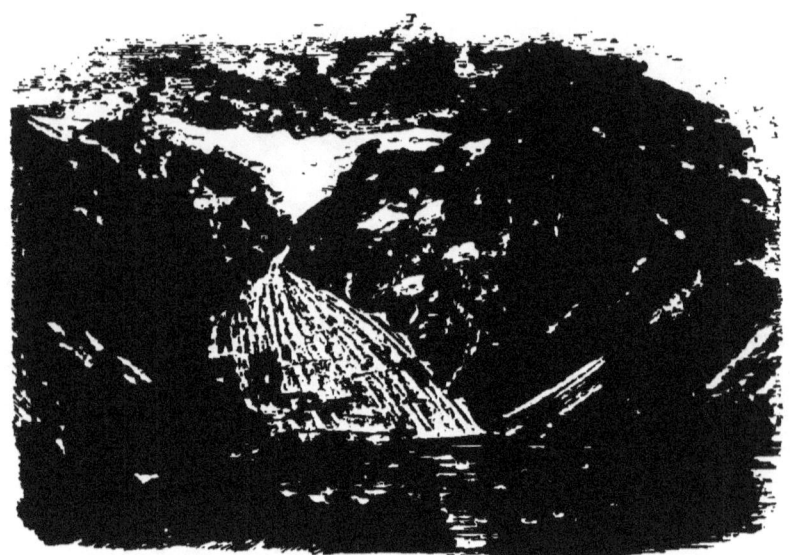

Fig. 70.—Glacier descending to the sea. Head of Jokuls Fjord, Arctic Norway.

27. We conclude then that one important part of the work of glaciers is the **transport of the materials of the mountains** from higher to lower levels. In the case of such a mountain chain as the Alps or the Himalaya, the glaciers melt long before they can reach the outskirts of the high grounds. They, therefore, do not bear their burden beyond the mountain region, though, as we have just seen, they once carried it much farther than they do now. But in Arctic and Antarctic regions the glaciers actually reach the sea-level, and, even pushing their way out to sea, break off into icebergs (Lesson XVI.). The accompanying drawing (Fig. 70) represents the little glacier at the head of the Jokuls Fjord in the north of Norway, which descends into the sea. A few transverse crevasses may be observed in it near the base. From time to time the outer portions break away, and fragments of ice, which are true miniature icebergs, float slowly on the current of fresh water moving down the inlet. Pieces of stone may now and then be seen upon these little bergs. In Greenland vast glaciers, like the Humboldt glacier, which measures 60 miles in width, descend to the sea-level, and extend for some distance from the shore, till large sections of their seaward ends split off and float away as icebergs (Fig. 16). Occasional blocks of stone have been noticed on icebergs on the ocean. The debris of the mountains is thus actually borne by the ice out to sea, and may travel for many hundreds of miles before, as the bergs melt, it sinks at last to the bottom of the ocean.

28. *Erosion.*—In the second place, a little further observation will suffice to show that the carrying of materials down its valley is not all the work which the glacier performs. Let us consider the river of muddy water that escapes from the end of the ice. Muddy at every season of the year, it is most copious and discoloured during the warm dry weather of summer and autumn. The mud cannot come from the melting of the glacier itself, for the ice is clear and pure. Very little of it can be derived from the bright little brooks and torrents which rush down from the melting snows and the springs on either side of the valley. Yet it undoubtedly comes out from beneath the glacier. Mud consists merely of the finer particles worn from rocks. There must, therefore, be a great deal of waste somewhere to account for this constant and plentiful supply of mud.

29. Following this inquiry further, we observe that the rocks which rise on either side of the valley from under the glacier are remarkably smoothed. Underneath the ice, as we may sometimes detect, the floor of the valley is similarly smoothed and polished. The contrast between the rounded and smoothed outlines of the knobs and hummocks of rock near the ice, and the sharp, rugged forms of the crags above, is often singularly well-defined. Besides this smoothing and polishing, the surface of the rocks may be observed to be covered with many parallel or intersecting scratches and groovings, varying from such lines as might have been graven by a hard grain of sand to such deep ruts as would require the forcible pressure of some sharp edge or blunt corner of stone. These markings are seen to run in a general direction down the valley. They have been evidently produced by some agent which has descended with sufficient force and steadiness to grind the rocks along the bottom and the lower slopes of the valley.

30. It is now and then possible to creep in under the ice at the end of the glacier, and to see where it rests upon its rocky bed. At such times we may, as it were, catch the glacier in the very act of grinding down and striating the rocks below it. Pieces of stone and grains of sand are jammed between the ice and the rock over which it moves. Held there, and pressed against the rock, they groove and scratch it. As this goes on year after year, the surface of the rock necessarily undergoes continual waste, and acquires that smoothed, polished, and striated appearance so characteristic of the bottom and sides of glacier-valleys.

31. Here, then, is the main cause of the unfailing muddiness of the water that issues from the end of a glacier. The glacier is busily engaged in wearing down its channel with the same kind of grinding powder which a river uses; but employs these materials in its own way, and forms with them a peculiar smoothed and grooved surface, such as no other agent in nature can produce. The earth, sand, and stones that fall from the moraine heaps through the crevasses, or between the glacier and its rocky sides, form the hard materials by which the ice erodes its channel. While employed to wear down the solid rocks, they themselves are undergoing constant wear and tear. They become smoothed, polished, and striated like the solid rocks, over which they are ground along. One of these ice-smoothed and striated stones is represented in Fig. 71.

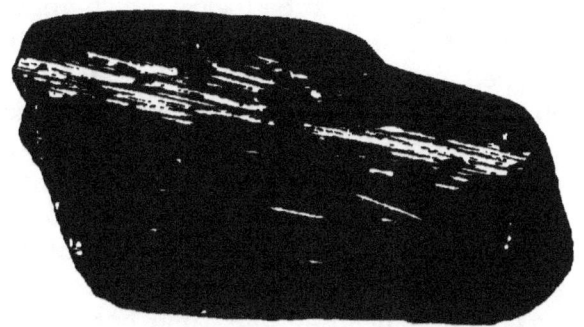

Fig. 71.—Stone polished and striated under glacier-ice.

32. We conclude, then, that another great task in which every glacier is ceaselessly engaged is the **erosion** of the sides and bottom of its valley. That this is an important work in regard to the general scenery of a country may be shown by the great height and the long distances to which the peculiar forms of ice-worn rocks may be traced. The former greater thickness and wider extent of the Alpine glaciers are not more decisively shown by the dispersion of the erratic blocks than by the range of the polished and striated rock-surfaces. By this test it can be proved that a great part of Northern Europe and America has been under moving sheets of land-ice, which, passing over the land from mountain-ridge to seashore, have left behind them their

memorial in the almost indelible markings engraved upon the rocks.

33. Again, no one can see the tumultuous body of discoloured water escape from a glacier without appreciating that in time a sensible deepening of the valley must take place in consequence of this ceaseless erosion and removal of materials. The deepening cannot be uniformly spread over the whole of the valley. There are places where the ice exerts more grinding power than at others, as a river at its falls and rapids effects more destruction than among meadows and plains. The rocks, too, of the glacier's bed vary much in hardness and power of resistance. Some parts must be more readily scooped out than others. So that, should the glacier retreat up the valley, these more deeply excavated portions, unless concealed under moraine rubbish, would become basins filled with water. Lake-basins of this kind in the midst of ice-worn rocks are a marked feature of glacier districts, and of all those regions of Northern Europe and Northern America which have just been referred to as having been at one time buried under ice (Lesson XXVI. Art. 4).

LESSON XXIX.—*The Sculpture of the Land.*

1. In Lesson XX. the leading features of the general external form of the land were described—its mountains, table-lands, valleys, and plains. At the end of that Lesson the question naturally arose whether any explanation could be given of the origin and history of these various features, but the answer to this question was postponed until after some consideration had been given to the nature of the materials and the internal constitution of the globe, and to the operations of water upon the surface of the land. We are now, therefore, in a position to return to the subject, and to apply to

the investigation of it the facts and deductions about the interior and exterior of the land which have come before us in the last eight Lessons.

2. The existing land consists of the higher parts, projecting above sea-level, of those ridges into which the exterior of our planet has been wrinkled during its gradual consolidation and contraction from an original fluid or viscous condition (Lesson XXII. Art. 20). We must not hastily conclude, however, that the land, such as we now see it, is the original surface of the solid globe. That it cannot be so must be evident from two considerations, to which reference has already been made. 1st. All the land, as far as we know, has been under the sea, and, even up to its mountain tops, consists in great part of hardened and altered sand, mud, and other materials, which were originally laid down upon the floor of the sea (Lesson XXI. Art. 13). 2d. The whole surface of the land is subject to a constant and enormous, though unequally distributed decay and removal. A comparatively short period would suffice for the entire destruction of the continents, were their surfaces to be continuously wasted even at the rate of the Mississippi's operations—and other rivers work considerably faster. At any probable rate of degradation, the land surface that first appeared above the earliest ocean must have been long since destroyed, and we can hardly hope to find any trace of it, even buried under the later accumulations of which the continents consist (Lesson XXVII. Art. 22).

3. But though no portion of the present land can be looked upon as part of the original or earliest solid surface of the planet, there can be no doubt that the existing continents must be very old. Not improbably they occupy the sites of the first ridges that appeared upon the cooling and shrinking mass of the globe. In the course of ages, these primeval ridges would be worn down by the action of water and air. But from time to

time, if renewed uprisings took place along the same original lines, the land would be formed and destroyed, and then formed and destroyed again. That this view is not mere theory, but rests on a strong basis of probability, may be shown by a consideration of the way in which the materials forming the land have actually been put together. If the existing ridges can be proved to have been upraised again and again during past ages, they may at least be plausibly conjectured to mark generally the primeval lines of elevation on the surface of the globe.

FIG. 72.—Quarry in flat stratified rocks.

4. Let us, then, return to the composition of the earth described in Lesson XXI. All over the globe, it is found that by far the largest mass of the land is built up of materials that have been slowly accumulated as sediment on the floor of the sea. These materials are arranged in layers or strata which have been laid down upon each other, until a depth of many thousands of feet has been formed. It is evident that the original position of these strata must have been nearly or quite horizontal, seeing that they were piled one upon another, as sand and mud are laid down on the level or gently slop-

ing bed of the sea at the present time. The subterranean movements whereby they have been raised above the sea into dry land, have taken place over such wide regions that this original level or gently inclined position may remain. The flat bedding of the rocks shown in Fig. 26 is of common occurrence, and most people will recognise the familiar look of such sections as that of the quarry represented in Fig. 72. In such cases of horizontality, we may conceive a large tract of the sea-floor to have been raised up into land so uniformly and equably, that the sheets of hardened sand and mud remained nearly, or quite in their original level condition. In Central and Northern Russia, in China, and in the Western Territories of the United States, this gentle and equable upheaval has taken place over regions many thousands of square miles in extent.

5. While the earth has been contracting as a result of its cooling, the effects of this contraction have not been uniformly distributed over the surface. The vast basins of the oceans, no doubt, mark the regions where the subsidence has been greatest. Probably they have been depressions from the beginning, though portions of them, particularly along their margins, have from time to time alternately risen and sunk. Every tract which sinks requires to accommodate itself to a diminished superficial area, and therefore exerts a strong lateral thrust upon the adjoining more stable parts. Under the influence of this force, long ridges have been raised into land between the ocean basins. Every successive subsidence may thus have carried with it a corresponding upheaval. So that, as regards the whole globe, although subsidence was the rule, and although the land was being continually wasted by air, rain, rivers, and the sea, nevertheless these periodic uplifts along the same general lines of movement, that is, along the axes of the continents, have compensated for the loss, and seem to have maintained, on the whole, the balance of dry land.

6. But it could hardly happen, in the midst of these movements, that the upheaval should be always so gentle and uniform as not to disturb the original level, or nearly level, position of the strata. On the contrary, over wide tracts of land, and more particularly along vast extended lines, the rocks have not only been upraised, but have been crumpled up and broken. Instead of remaining in horizontal or slightly inclined sheets, they may be found tilted up in all directions, and often placed on end like books on a library-shelf. Every great mountain-chain furnishes examples of these more complex arrangements. On the plains and lowlands, the rocks may stretch for hundreds of miles as level as before they rose out of the sea. But towards the interior they begin to bend in wave-like undulations, which increase in magnitude until, along the flanks of the mountains, the rocks are sometimes found actually so thrown over that the lowest lie uppermost. This structure is explained in the accompanying diagram (Fig. 73).

7. In the form of mountain-structure, illustrated in Fig. 73, there is evidence of only one general upheaval which took place after the formation of the various rocks of the region, for these have participated in the change of position. In the example represented in Fig. 74, there is proof of two upheavals. First, the older series of rocks, A, was contorted and raised ; then, against its sides and upon its broken and worn edges, the series B was formed, and thereafter raised into land. A still

FIG. 73.—Section across a mountain-chain to show how the level rocks of the plains are bent and inverted along the flanks of the mountain, while the lowest and oldest rocks are made to form the central and highest point of the chain.

longer succession of movements is shown in Fig. 75. We there see that after the two movements which disturbed the rocks A and B, a third uplift took place, whereby a still newer series, C, which had been deposited upon the side of B, came to be raised into land.

8. Two facts about mountains are presented to the mind by such sections as these. First, that a great axis of elevation on the earth's surface, in other words, a mountain chain, may have again and again served as a line of relief from the strain of terrestrial contraction, and may have consequently been successively pushed up between subsiding areas on either side. And secondly,

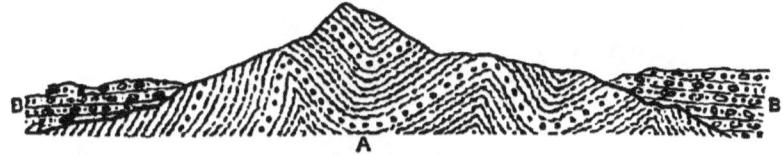

Fig. 74.—Section across a mountain-chain, showing two successive periods of uplift.

that between each period of uplift there has been great waste of the upraised land—air, rain, frosts, brooks, rivers, glaciers, and the sea, all wearing down the surface and producing the materials out of which the next series of rocks was formed.

9. It is evident also that by means of such sections we may compare the relative dates of different mountain-chains. If the rocks represented in Fig. 73 be the same as those marked C in Fig. 75, then the mountain shown in the former section must be much younger than that shown in the latter. We see, indeed, that the mountain in Fig. 75 is not only older, but older by two earlier upheavals. It is the province of the science of Geology to investigate these questions regarding the structure and age of mountain-chains. Geologists have discovered that among the rocks of the earth's crust a chronological order can be made out, and they can thus assign to each

great elevation on the surface of the globe its relative date in the history of mountains.

10. But while the existence of dry land and of the mountain-chains which traverse it must be ascribed to movements of the solid crust of the earth, the present aspect of the surface of the land cannot be its original aspect, but must have been largely determined by the action of those various agents which wear down its surface (Art. 2). The vast amount of mud annually transported into the sea by rivers proves how much material is continually removed from the land, and therefore, how

Fig. 75.—Section across a mountain-chain, showing three successive periods of uplift.

greatly, though it may be insensibly, the height and appearance of the mountains and valleys must, in course of time, be changed. This progress of disintegration and removal is going on all over the globe, here more rapidly, there more slowly, but always advancing, and always involving changes in the aspect of the land. In the lapse of the long periods of time during which it has been in progress, how vast must have been the mutations on the surface of the globe, how many successive mountain ranges may have been upraised and worn away !

11. The working of the chief agents that erode the surface of the land has been described in previous lessons —the air, by its gases and vapours, its winds and changes of temperature; frost by its oft-renewed wedges of ice; rain, brooks, and rivers, by their movement down the land, and their power to sweep along the loosened debris; avalanches and glaciers, by the fragments of stone with

which they grind and polish the rocks of the valleys; the sea, by the waves thrown incessantly against its shores. We may compare the general results of the co-operation of all these forces to the work of a sculptor. They are, so to speak, the different tools with which the framework of the land is carved. But the sculpture they achieve is never completed. It goes on continuously so long as the land remains above the sea.

12. At first it may seem almost incredible that the whole surface of the land, even the loftiest and stateliest mountains, should thus be crumbling down. But the more we search for proofs of the assertion, the more clear and abundant do they become. We learn that, whatever may have been the aspect of the land when first pushed out of the sea, it has been, and is now being, chiselled from its highest peaks down to below the tide-marks. Its cliffs and pinnacles are split up and grow more shattered and sharp every year. Its ravines are widened and deepened. Its hilly surfaces become more roughened and more deeply seamed by the lines which running water traces over them. Its valleys and plains are levelled and strewn with debris washed down from higher grounds.

13. In travelling from place to place we cannot fail to notice the evidence of this universal decay, and, on further observation, to remark that, though the wearing down of the land may be traced more or less clearly everywhere, its rate and the changes of scenery which it brings with it depend very much upon the nature of the rocks of each region. Here again we may have recourse to the simile of the sculptor's work. The character of a statue depends not only on the design and manipulation of the artist, but also on the material employed. Out of a piece of granite or of pudding-stone, no matter what amount of genius and skill were bestowed on it, the same effect could never be produced as from a block of white marble. So we find that the hills, valleys, and mountains differ from each other, in great measure, according to

Y

the nature of the rocks of which they consist. In our journeys, whether in our own district or in other regions, we shall find it not uninteresting nor uninstructive to take note of the changes in the aspect of the scenery through which we pass, and to connect them with variations in the character of the rocks.

FIG. 76.—Scene on the Coast of Caithness.—Influence of joints among stratified rocks in the formation of vertical cliffs and outstanding square blocks of rock.

14. Apart from the varying nature of the materials, nothing contributes more to the character of scenery than those lines of division or "joints" which have already been referred to as traversing all rocks. They serve as channels for the descent and reascent of sub-

xxix.] THE SCULPTURE OF THE LAND. 323

terranean water (Lesson XXIV. Art. 17). They are

Fig. 77.—Portion of the west front of Salisbury Crag, Edinburgh, showing the influence of joints in promoting the splitting up of the igneous rock, and the preservation of a vertical face to the cliff.

made use of by frost as the lines along which the wedges

of ice are driven most effectively into exposed faces of rock (Lesson XXVIII. Art. 10). Every mountain-peak and cliff, indeed, every large naked mass of rock which projects into the air, owes more or less of its characteristic outlines to the way in which its joints have been split open. Among the stratified rocks the joints allow vertical cliffs to be formed and large square buttress-like masses to project from the cliff or even to be isolated from them. The foregoing drawing (Fig. 76), for example, shows how the forms of coast-cliffs are determined by the position of the intersecting lines of joint, each vertical face of rock corresponding with the direction of one of these lines. Where such rocks form lofty mountainous ground, they are often carved into the most picturesque forms of pinnacle and buttress. Again, among the unstratified rocks, such as granite and basalt, the influence of the joints is no less marked, as, for instance, where it defines the ledges and rifts in a precipice, or where it has allowed the most solid rock to be so completely shattered as to look like a huge mass of ruin. In Fig. 77 a representation is given of part of the face of a basalt cliff, where abundant joints traverse the rock in such a way as to divide it into rude prisms, which are gradually wedged off from each other by frost, until, detached at last, they fall to the base of the precipice. The ruined masses below are further broken up and carried away piecemeal by frost, rain, and in the general process of "weathering," or in other cases, by waves along a sea-shore or by the flow of a brook or river. Their removal permits the continuous decay of the cliff and preserves the steep face of the precipice, which thus slowly recedes as slice after slice is cut away from its front. But where the detached blocks gather at the base, they form in the end a protecting bulwark, and either retard or prevent the farther recession of the line of cliff.

15. Among the higher parts of the mountains this

kind of rude chiselling of the rocks is most conspicuous. Not only are steep crags and lofty precipices formed, but the very mountain ridges are cut away into sharp crests. These, again, still further split and splintered by the severe frosts and furious storms of the mountain climate, are cut into slender pinnacles and spires, sometimes at a distance seeming so needle-like in their slimness and sharpness, as to have received, among the Alps, the name of *aiguilles*, or needles. The blocks loosened from these high crags and crests furnish abundant stones for glacier moraines (Lesson XXVIII. Art. 22).

16. From the top of a high hill, or of a mountain beneath the snow-line, one may sometimes look down upon a wide region and mark, as in a vast map or model, how the little gullies on the sides of the slopes widen out into larger channels, how these run together into valleys, and how the whole landscape seems thus to be trenched with water-courses. One who has the good fortune to see such a scene as this, after having learned to appreciate how ceaselessly and potently every brook and river is cutting out its channel (Lesson XXVI.), will realise more vividly than from any map or description how the valleys are carved out by the power of running water. Every little gully and ravine down the steeper declivities is a sample of the way in which the forms of the solid land are changed; every gradation of size and shape may be traced, from the trench that was opened by some storm last winter, up to the deep and wide gorge through which a foaming river rushes, or the ample valley down which the collected waters from a whole range of mountains sweep onward to the sea. One may not be able to tell how far the line of any particular valley may have been originally determined by the shape which the ground had when it first rose out of the sea. But, as a heavy shower of rain produces runnels which soon cut out a miniature drainage system on a roadway, so, in the course of time, the flow of brooks and rivers over

the surface of the land must necessarily erode systems of valleys. Whatever might have been the original shape of a country, rain and frost, brooks and rivers, snow-fields and glaciers could not have been at work upon it for even a comparatively short time, without carving out valleys for themselves, and sculpturing the mountains into such sharp and rugged forms as they now wear. Everywhere there lie proofs of excavation; hillsides are furrowed with gulleys, mountain slopes are trenched with ravines, tablelands are cut down until they become only chains of ridges separating the valleys which have been carved out of them (Lesson XX. Art. 19).

17. While the effects of disintegration in roughening the surface of the land are most marked among the high grounds, the results of this process show themselves among the lower regions in the strewing of the crumbled fragments of the hills over the valleys and plains. Every tract of fertile meadow or level field bears witness to the way in which the lowlands have been smoothed and raised in level by the sand and earth spread over them by rain and streams. And yet this increase of height in the plains does not really compensate for the waste of the high-grounds. A little consideration of the matter shows, in the first place, that though the plains do obtain considerable additions to their surface from the materials swept down upon them by rivers, they receive only a part of these materials, the remainder being carried out to sea; and in the second place, that even the plains themselves are wasted; floods tear up their soil and sweep away their river-banks.

18. It appears, then, that the tendency of the process of sculpture, which gives to landscapes their characteristic details of outline, is in the end to reduce the dry land to the level of the sea. But this is not all. While the general surface of the land undergoes attacks from the atmospheric influences its margin is continually

suffering from the assaults of the waves. Only the parts of the earth's surface lying under a considerable depth of ocean are protected from decay (Lesson XVIII. Art. 19). Along the margin of the land the waves are gnawing away the coast-line, or are only kept from doing so by the bar of detritus which has been thrown up against them.

19. Were no other operation to come into play, the natural and inevitable result of this ceaseless destruction would be the final disappearance of dry land. But here we see the meaning and importance of the underground movements already referred to as the results of terrestrial contraction. The great ocean basins have from time to time sunk down, and in so doing have pushed up ridges of land between them. These ridges, on each successive uplift from beneath sea-level, consisted mainly of the more or less consolidated debris worn away from their predecessors. The same materials have thus served, over and over again, to form the framework of the upraised land, and thus, while looking at the subject from one side, we see only ceaseless destruction, mountain and valley continually crumbling down before us; yet, taking a wider view, we perceive that decay of the surface is needed to furnish soil for the support of living plants and animals which people the earth, and that the materials so removed from sight are not lost, but are carefully stored away, to be, in some future time, raised into new land, thereafter to go again through a similar cycle of change.

CHAPTER V.

LIFE.

LESSON XXX.—*The Geographical Distribution of Plants and Animals.*

1. The foregoing Lessons have treated of the parts of the earth, their relation to each other, and the constant changes and reactions between them which constitute the Life of the globe. But above and apart from all these movements within or upon the surface of the earth, another kind of activity and progress now claims our attention, where the forces concerned are not air, sea, and land, but the living energy of plants and animals. Our planet is not merely a theatre for the evolution of physical phenomena. It has been appointed as the dwelling-place of a vast and varied series of living things, which move through the air and people both land and water. The study of these living organisms is comprised under the general name of Biology, or the science which deals with vegetable and animal life.

2. So vast a study, opening up wide fields of inquiry far beyond those over which we have been travelling in these Lessons, must needs be subdivided into different departments. Thus one branch inquires into the structure and growth of plants, another deals with the way in which plants are distributed over the globe, a third treats of the structure of the various tribes of animals, a fourth follows the action of the different parts of an animal's body and the part which each of these plays in the life

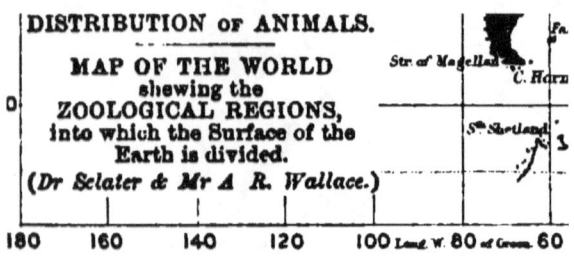

of that body, a fifth arranges the enormous numbers of animal forms in due order, to show their grade in the scale of being, and to allow the general assemblage of living forms in one region to be compared with that in another. These and the other branches of biology deal chiefly with the plants and animals as they are in themselves, or as they stand in relation to each other.

3. But it is evident that, just as we ourselves are encompassed by external conditions of geography, climate, and vegetation which, it may be unconsciously, govern our everyday life, so each plant and animal on the globe comes under the control of similar surrounding influences. Apart, therefore, from the structure, functions, or classification of vegetable and animal life, we may study it with reference to its relation to these external and dominant conditions. From this point of view, physical geography and biology are seen to be closely linked to each other. It is impossible to gain any intelligent conception of the present distribution of plants and animals over the globe without entering upon inquiries which form part of the scope of physical geography.

4. We all know how greatly the plants and animals of different quarters of the globe differ from each other. The equatorial regions nourish a rank and luxuriant vegetation, including palms, bananas, tall tree-grasses with rope-like lianas twisting round their stems, and bright-hued, strangely-shaped orchids hanging from their branches. The animals are equally characteristic, for they include lions, tigers, elephants, rhinoceroses, camels, giraffes, crocodiles, large serpents, with crowds of gorgeously-plumed birds and brilliant butterflies.

5. The temperate zone is distinguished by a very different assemblage of plants and animals. The forests and woods show such trees as the oak, ash, elm, sycamore, beech, poplar, birch, hazel, and pine. The dells are bright in spring with snowdrops, anemones, and primroses; and in summer with speedwells, geraniums,

and wild roses. But both the plants and the animals are more sober in colouring than in the hotter parts of the earth. The birds include thrushes, larks, and other songsters. Among the wild animals of the low-ground we find mice, rats, weasels, hedgehogs, badgers, otters, foxes; and in hilly districts wild-cats, wolves, and bears.

6. Within the Arctic and Polar regions life becomes much less abundant and varied. As we advance, trees disappear, though stunted forms of birch, fir, and willow extend a long way northwards. By degrees these too die out, and the scanty vegetation consists, at last, mainly of mosses and lichens, with saxifrages, gentians, and a few more flowering plants. These snow-covered lands are wandered over by polar bears, white foxes, reindeer, musk-oxen, and ermines; the seas are frequented by seals, walruses, and whales; while the coasts are sought by large flocks of northern sea-fowl and by snowy falcons, buntings, ptarmigans, owls, and other white-feathered birds. At the extreme northern limits reached by explorers, life of any kind is hardly to be met with among the deep snow-fields and piled-up heaps of ice which cover land and sea.

7. From the extreme exuberance and variety of plant and animal life in the equatorial and tropical lands there is thus a gradual diminution polewards, until in the far polar regions almost the zero of vegetation and of animal existence is reached. There can be no doubt, therefore, that one grand influence in the distribution of plants and animals is temperature; warmth being favourable, cold unfavourable, to the growth of living things.

8. But if temperature were the sole cause that determined the character of the plants and animals of any country, then every zone of latitude should be marked by the same kind of vegetation and by the same groups of animals. It would then be enough to know the geographical position of any place to be able to tell what must be its *flora*, that is, its assemblage of plants, and

its *fauna*, or population of animals. But a very little inquiry suffices to show that no such strict coincidence between latitude and the distribution of plants and animals really exists. The Old and the New Worlds are traversed by the same isothermal bands (Lesson IX. Art. 2), and have similar kinds of climate, soil, and exposure. Yet, in regions closely resembling each other as to conditions of physical geography, the plants and animals, though sometimes presenting such general resemblance as to show them to be mutually representative of each other, are often widely different in the two hemispheres. The lion of the Old World gives way to the puma in the New; the tiger is replaced by the jaguar; the elephant, rhinoceros, and hippopotamus, by tapirs and peccaries; camels by vicunas; apes and baboons by flat-nosed monkeys and marmosets. The birds are not less distinct; the Old-World vulture being represented by the New-World condor; sun-birds by humming-birds.

9. The present arrangement of plants and animals over the earth's surface is one of the most difficult questions with which science has to deal. The first step towards its investigation must necessarily be a careful inquiry into the actual facts regarding the distribution of the various kinds of plant and animal life. This inquiry has been carried on by naturalists in all parts of the world with the result of enabling them to parcel out the earth's surface into distinct regions and sub-regions, each characterised by a peculiar flora and fauna, though sometimes containing a number of the same plants and animals in common. These regions, to which the following names and limits have been assigned, are shown in Plate X.[1]

O. (i.) **Palæarctic Region**, comprising all Europe with the temperate parts of Asia, and the tracts of Africa

---

[1] The arrangement here followed is that of Mr. Sclater, as modified by Mr. Wallace in his admirable work on *The Geographical Distribution of Animals*.

lying to the north of the Sahara desert; or the northern parts of the Old World, from Iceland to Behring Strait, and from the Azores to Japan. In the Arctic portions of this vast region the vegetation is comparatively meagre, showing an abundance of mosses and lichens, which, in the tundras of Siberia, cover thousands of square miles of barren waste. In favourable places, flowering plants, such as saxifrages and gentians, peep out from beneath the snows during the short but warm summer; stunted forms of willow, azalea, and rhododendron form here and there a kind of scrub upon the slopes, while southwards from the mean annual isotherm of 32°, pine-trees make their appearance and increase in number, till they form wide ranges of dark forest, as in Norway, and on the higher mountain groups farther south. Between the isotherms of 40° and 60° the sombre pine-forests, retaining their leaves throughout the year, give place to a more varied and luxuriant *deciduous* vegetation, which sheds its leaves in autumn and renews them again in spring. The trees include many noble kinds,—birch, alder, beech, ash, oak, elm, sycamore, walnut, chestnut, and maple. Northern fruits, like the cranberry, cloudberry, bilberry, strawberry, currant, and raspberry, are succeeded farther south by luscious pears and apples, almonds, olives, figs, grapes, and oranges. The cereals —wheat, barley, oats, etc.—are abundantly cultivated throughout most of the region. In the more southern countries, with a mean annual temperature, between 60° and 70°, the trees do not lose their leaves in winter. Here we find the evergreen oak, myrtle, and laurel, with some plants, such as palms, which more properly belong to the tracts lying nearer the equator.

11. The animals of the region show a development similarly connected with the distribution of temperature. In the extreme north, white bears and foxes, reindeer, whales, walruses and seals, many peculiar sea-birds, together with white owls and ptarmigan, form a distinct

assemblage. As most of these species disappear southwards, their places are taken by brown bears, badgers, otters, horses, buffaloes, fallow and roe-deer, chamois, wild goats, wild sheep, hares, rabbits, moles, hedgehogs, and dormice; golden eagles, hawks, grouse, pheasants, house-sparrows, magpies, jays, and thrushes. Farther south the camel is the most characteristic animal.

12. (ii.) **The Ethiopian Region** embraces Central and Southern Africa, with the tropical part of Arabia, Madagascar, and the neighbouring islands. To the south of the great desert of Sahara the western portion of the African continent is largely covered with dense forests, where the oil-palm, the huge baobab, euphorbias, bignonias, tamarinds, and many other tropical plants, form a dense luxuriant vegetation in the hot moist air of that climate. Farther to the east lie vast elevated lands, covered with tall grasses and sedges, and dotted with patches of forest. The flora of the southern part of Africa is distinguished by its great variety of heaths, its fig-marigolds, carrion-scented stapelias, aloes, and pelargoniums.

13. The fauna of the Ethiopian region is marked on the one hand by the absence of such wide-spread animals as camels, deer, goats, sheep, wild oxen, wild boars, and bears; and, on the other hand, by the presence of many remarkable forms of life, including the gorilla, chimpanzee, baboon, lemurs, aye-aye, lion, leopard, civet, hyæna, zebra, rhinoceros, hippopotamus, giraffe, antelope, elephant, ostrich, ibis, flamingo, chameleon, and crocodile.

14. (iii.) **The Oriental Region** includes Southern Asia, from the mouth of the Indus along the southward slopes of the Himalaya mountains and the Chinese uplands to Ningpo, with Formosa, the Philippines, Borneo, and the Malay Islands as far as the south-east end of Java. Much of the surface of this region is covered with dense forests of tropical vegetation. Among the better-known plants occur the ginger, arrow-root, banana,

cocoa-nut, screw-pine, yam, bamboo, rice, gourd, custard-apple, mango, coffee-tree, mangrove, ebony-tree, bignonia, hemp, and sandalwood.

15. The fauna contains some characteristic animals—the ourang-utan, long-armed monkeys, flying lemur, many civets, the tiger, hyæna, jackal, wild cattle, elephant, rhinoceros; many bright-feathered birds, as trogons, hornbills, goat-suckers, sunbirds, long-tailed parrots, and peacocks; numerous reptiles, including ground and tree-snakes, cobras, and crocodiles; and a vast assemblage of insects, among which the size and brilliancy of the butterflies and many of the beetles are remarkable.

16. (iv.) **The Australian Region**, embracing Australia, New Zealand, and the numerous islands to the east of Java, Borneo, and the Philippine group, consists wholly of islands, which, lying apart from all the great continental masses of land, show a peculiar assemblage of plants and animals. The vast insular expanse of Australia, situated partly within and partly without the tropics, and exposing a wide desert interior to the hot rays of the sun, while its coast-line is washed by the open sea, presents contrasts of climate not met with in the smaller islands of the region. From its size, also, and its proximity to the south-eastern limits of the Oriental region, it contains a greater diversity in its flora and fauna. Over the dry and warm tracts of Australia, the general heath-like vegetation is marked by a pervading dead blue-green colour, with dull leaves so arranged upon the plants as to afford but little shade. The eucalyptus, or gum-tree, and other trees and shrubs bearing bright honeyed flowers, together with thickets of acacia and scattered marsh-oaks, give a peculiar character to the forest-lands. Vast regions are covered with grasses, and furnish good pasture. Along the northern limits, where this region borders the Oriental islands, the flora contains some of the forms of vegetation found more plentifully towards the north and north-west, such

as pandanus, cabbage-palm, fig, nutmeg, and sandalwood. On its southern margin, where the climate assumes a moister and more temperate character, ferns, cycads, and pines abound, while the heath-like epacris, and numerous proteas enliven the surface with their bright blossoms. New Zealand is distinguished by the verdure of its flora, which consists largely of ferns, often growing as trees, and many kinds of pines. The scattered islands of the Pacific have their cocoa-nut palm, breadfruit, tacca, grasses, and sedges.

17. The fauna of the Australian region is the most peculiar on the face of the globe, both for the types which it contains, and for the almost universally diffused forms which it has not. No apes or monkeys chatter in its woods, no wild horses, cattle, or sheep browse over its pastures, no wolves, foxes, tigers, or other similar beasts of prey prowl across its hills and valleys. The place of these various and wide-spread species is taken in Australia by a totally distinct and less highly-organised class of animals called marsupials, of which the kangaroo may serve as the type. Of these there are many varieties, some living on fruits, others on roots, others on smaller animals of their own kind, others on insects; some keeping to the ground, others taking to trees. The birds likewise are peculiar, for they include the bird-of-paradise, lyre-bird, cassowary, paroquets, and honey-suckers.

18. (v.) **Neotropical Region.**—Under this term are included the whole of South America, the islands of the Antilles group, and the tropical part of North America. Ranging across the whole zone of the tropics, and as far south as the 56th parallel of south latitude, and rising up to the snow-line in the Andes, this region presents many varieties of climate, which are well shown by differences of vegetation. The lower grounds within the tropics present the most luxuriant flora in the world. It abounds in mangroves, palms (cabbage-palm, ivory-palm, and

other kinds), bananas, tree-ferns, and mimosas, growing in dense jungles, and having their stems and branches clustered round with many smaller plants, such as lianas and ferns, or gorgeously-blossomed cactuses, orchids, and passion-flowers. The vast plains or *llanos* of the Orinoco, with their tall grasses and occasional clumps of pines and mimosas, represent the pasture lands of the Old World, but with their bright lilies show a brilliancy of colour peculiar to themselves. Farther south the basin of the La Plata presents similar plains or *pampas*, which, getting less and less luxuriant in their vegetation, are at length succeeded by the barren moors of Patagonia and Tierra del Fuego. On the lower mountain slopes, characteristic trees are cinchonas, from the bark of which quinine is prepared. Mahogany, rosewood, the indiarubber tree, with many plants yielding spices, balsams, and perfumes, give a distinctive character to the South American flora. On the more elevated tracts, calceolarias, gentians, and low-growing plants occur, that remind the traveller of some features of mountain vegetation near the snow-line in the Old World.

19. The fauna of this region is more varied than that of any other. It contains the peculiar jaguar, flat-nosed monkeys and marmosets, blood-sucking bats, chinchillas, sloths, armadilloes, ant-eaters, racoons, opossums, deer, llamas, alpacas, vicunas, tapirs, and peccaries; but no native sheep or oxen. Among the birds occur condors, curassows, rheas, or American ostriches, toucans, jacamars, mot-mots, macaws, and numerous forms of humming-bird. The reptiles include the boa-constrictor and many other serpents, the alligator, crocodiles, tortoises, and turtles. The insect life is immensely abundant and varied.

20. (vi.) **The Nearctic Region** embraces all North America lying to the north of the tropic of Cancer. Its greatest breadth lies towards the cold northern regions; whence it rapidly narrows southward, so as to be con-

nected with the Neotropical region merely by a narrow strip of land. This isolation is accompanied by a somewhat less varied flora and fauna than in the corresponding region of the Old World. The plants and animals, taken as a whole, present much less contrast to those of the Palæarctic region than those of the Neotropical and Ethiopian regions do to each other. The northern tracts of North America extend far within the Arctic Circle, into the snow-covered lands where vegetation reaches its lowest point of development. The southern limits of the province, on the other hand, lie towards the tropical zone, where the sugar-cane, yucca, cotton, maize, and tobacco are characteristic plants. In California and Oregon many large and distinct kinds of pine occur in the forests, such as the gigantic Sequoia (Wellingtonia) and the Douglas pine. Eastward of the Rocky Mountains vast undulating pasture lands or prairies stretch over the basin of the Mississippi and its tributaries. The British Possessions are covered with extensive forests, which, as the country becomes peopled, are gradually giving way to pasture and cultivation.

21. The fauna varies with the latitude. In the north are found musk-sheep, moose-elks, reindeer, gluttons, skunks, racoons, beavers, lemmings, jumping-mice, and tree-porcupines. Farther south vast herds of bisons roam over the prairies. Other typical animals are the grizzly bear, black bear, puma, lynx, prong-horned antelope, prairie-dog, flying-squirrel, pouched-rat, opossum, humming-bird, blue-crow, and rattlesnake.

LESSON XXXI.—*The Diffusion of Plants and Animals.*

Climate.—Migration and Transport.—Changes of Land and Sea.

1. It is not enough to know how the various tribes of

plants and animals are distributed over the surface of the globe. We are irresistibly led to ask ourselves why and how their distribution has come to be as we have traced it in the preceding lesson. Not very long ago, men were content with supposing that the present arrangement had always existed, ever since the different continents and islands rose out of the sea and received their earliest inhabitants.

2. But in the gravels and clays beneath the soil, or in the limestones, sandstones, and other rocks lying below, traces have been found all over the world of older and different plants and animals, which occupied the land before the present denizens had appeared. Our modern horses and cattle, for example, were preceded by other kinds which are no longer living. The bears, wolves, and hyænas of to-day are not quite the same as those of which the teeth and bones are found in caves and ancient alluvial deposits.

3. A thoughtful study of this subject suggests three main influences which may have guided the distribution of the present faunas and floras of the earth's surface. 1st. Climate. 2d. Migration and transport. 3d. Changes in the form and height of the land and in the depth and extent of the seas. A knowledge of the nature and effect of these influences helps us to understand much that would otherwise be inexplicable, but there still remain, and perhaps must ever remain, many difficulties which no amount of research may be able to remove.

I.—Climate.

4. This term includes the general temperature, moisture, winds, and other atmospheric conditions which prevail in any district, and which directly affect the growth and vigour of plants and animals. From the statements made in Lesson IX. it appeared that the

climates of different portions of the globe greatly differ, and some of the causes of such differences were there traced. But a knowledge of the annual distribution of heat at any place, though it gives us one main element in determining the climate of that place, requires to be enlarged by further knowledge respecting the rainfall, the direction of the prevalent winds, the shape, height, and position of the ground, the character of the soil, vegetation, and other more local details. Since Lesson IX. we have been led over most of these subjects, so that we may now return to the question of climate in reference to the arrangement of plants and animals.

5. On due consideration of this subject, five distinct influences by which the climate of any place is determined may be recognised. 1st. Distance from the equator. 2d. Distance from the sea. 3d. Height above the sea. 4th. Prevailing winds; and 5th. Local influences, such as soil, vegetation, and proximity to lakes or to mountains. Each of these causes directly, and often powerfully, controls the spread of vegetable and animal life.

6. (i.). **Distance from the Equator.**—Climate, having temperature for its main element, must follow generally the course of the isothermal bands over the earth's surface. The warmest climates are necessarily those of the inter-tropical regions, where the sun's rays are vertical, or not much inclined from vertical. In proportion as we recede from the equator, the rays fall more and more obliquely, and the same amount of heat-rays is therefore spread over an increasing breadth of surface, while, moreover, they have to pierce a greater mass of air. Round the poles, the least amount of heat is received, and the climates are coldest.

7. Were the earth's whole surface either land or water, the climates would be arranged in parallel and regular bands from the equator to the poles. But owing to the way in which land and water are grouped, such an

arrangement has been prevented (Lesson IX.). Two places on the same latitude may have very different average temperatures, and therefore very dissimilar climates. Nevertheless, the great predominating influence of position with regard to the equator is, on the whole, maintained in the arrangement of climates. It influences, in a marked way, vegetable growth, as is shown by the time of flowering or of ripening among widely-distributed plants. On the continents, this time becomes later in proportion as the country is distant from the equator. Thus the elm comes into leaf, at Naples about the beginning of February; at Paris, not until late in March; and in the centre of England, not until the middle of April. Ripe cherries may be gathered in the south of Italy about the beginning of May; they are ready in Northern France and Central Germany at the end of June; but not generally in England for three or four weeks later. Nothing could show more strikingly the difference of climates between the different latitudes of a continent.

8. But these differences are not merely marked by the variations in the growth of the same plants. As shown in last lesson, when we pass from one climate to another we encounter different plants and different animals. One by one characteristic forms of life drop away, and their places are taken by others. So constant and marked are these changes that such expressions as "an arctic vegetation," "a temperate flora," "a tropical fauna," have passed into general use, and convey a distinct picture to the mind.

9. We have found, however, that if such a distribution of plants and animals were due to differences of climate alone, wherever the same climate recurs it should be accompanied by the same kind of vegetation and of animal life, but that no general coincidence of this kind exists, when regions remote from each other are compared. The climate of Central Europe closely

resembles that of parts of the United States. Yet the wild animals and birds are strikingly different; mice, hedgehogs, buffaloes, chamois, and jays in the Old World are replaced by jumping-mice, racoons, opossums, bisons, llamas, and humming-birds in the New. In Central South America the forests are tenanted by jaguars, sloths, armadilloes, tapirs, curassows, and toucans. On corresponding latitudes in equatorial Africa these animals are represented by lions, leopards, hyænas, hippopotamuses, elephants, guinea-fowl, and touracoes. In Australia these forms are again replaced by a strange and peculiar assemblage of animals, including kangaroos, wombats, flying opossums, emus, lyre-birds, and crested pigeons. While, therefore, difference of latitude usually means difference of climate and of plant and animal life, identity of latitude with similarity of climate does not necessarily imply agreement in the character of the flora and fauna.

10. (ii.) **Distance from the Sea.**—The influence of the sea upon the distribution of temperature and moisture has been already described (Lessons IX., X., and XVIII.) As water is more slowly heated and cooled than land, the climates of the sea and of the coasts of the land are much more moist and equable than those of the interior of the land. In proportion, therefore, as places recede from the sea, their climates become more extreme. An insular or oceanic climate is one where the difference between summer and winter temperature is reduced to a minimum, and where there is a copious supply of moisture from the large water-surface. A continental climate is one where the summer is hot, the winter cold, and where the rainfall is comparatively slight.

11. These variations cannot but make themselves visible in the distribution of plant and animal life. They are well shown by contrasting the times of flowering and ripening of the same plants along the Atlantic border and in the central countries of Europe. It will be remem-

bered that owing to the influence of the warm Atlantic water the temperature of the whole of the north-west of that continent is raised considerably higher than it would otherwise be. (Lesson XVIII. Arts. 7, 8.) Consequently vegetation is much earlier in the south of Sweden than in the same latitudes to the east. The lilac and elm begin to show their leaves sooner at Upsala than at Paris, and while winter still reigns to the east of the Baltic, spring blossoms have already spread far up into Scandinavia.

12. The difference between an insular or oceanic and a continental climate is likewise well brought out by the fact that such evergreens as the Portugal laurel, aucuba, and laurustinus grow luxuriantly even in the north of Scotland, while they cannot withstand the severe cold of the winter at Lyons.

13. (iii.) Height above the Sea.—In Lesson IX. Art. 20 reference was made to the gradual diminution of temperature with increase of elevation above the sea. This cause of variation in climate is of a more local character than the two already illustrated. But its effects upon the spread of plants and animals is singularly well marked. If the fall of the thermometer be taken at 1° Fahr. for every 300 feet of ascent, we can readily perceive that the times of flowering and ripening of the same plants must become later, in proportion to the height of their place of growth, until at last the ground is too high and bleak to let them ripen at all before the winter sets in. They have thus an upper limit beyond which they cannot extend. But while they disappear, other plants, better able to withstand the rigorous climate of the uplands, take their place. As we climb to higher elevations the familiar vegetation of the plains is gradually succeeded by a vegetation peculiar to the mountains, until at last we reach the edge of the snow line. This influence of height on vegetation is illustrated in a graphical form by Fig. 78.

FIG. 78.—Vertical distribution of climate on mountains.

On the animal world, too, the influence of elevation may be distinctly seen. In Europe rabbits, moles, hedgehogs, otters, foxes, larks, thrushes, lapwings, and many other common forms occur among the lower grounds, while in the mountains such animals as the marmot, goat, ibex, chamois, brown bear, and eagle find their congenial home.

14. (iv.) Prevailing Winds.—Air lying upon the surface of any part of the globe tends to acquire the temperature of that surface. Consequently winds which come from a cold region are cold, those from a warm region are warm. Winds from the sea are usually moist, those from the land are generally dry. Sea-breezes are not liable to the same extremes of temperature as those from the land. The vapour which they carry with them cools the heat of summer and lessens the cold of winter. On the other hand, winds blowing from the interior of a continent are apt to be hot and suffocating in summer, piercingly cold and dry in winter. Winds which come from lower into higher latitudes, or from warmer to cooler climates, have their moisture condensed, and are therefore rainy, while those which blow from higher to lower latitudes, or from cold to warm regions, are dry.

15. Much, therefore, in the climate of any place, must be due to the prevailing winds. This is more particularly noticeable on the coasts of the continents, where the winds blow alternately from and to the sea. The striking contrasts between the extremely rainy and almost rainless districts in certain parts of India have already been referred to as showing the great influence of the winds in determining the moisture of a climate. (Lesson X. Arts. 32, 34.)

16. It is evident that, as regards plant-growth, moisture is hardly less important an element than temperature in the climate. Those tracts are most plentifully covered with vegetation which are most copiously watered. Both the abundance and the character of the vegetation depend greatly upon the amount of rain-fall. The west side of the British Islands, for example, which receives the first and largest precipitation of the moisture from the Atlantic, is much greener and more luxuriantly clothed with vegetation than the east side.

17. Whatever regulates the growth and distribution of plants must tell effectually upon the spread of animals. The herbivorous species naturally haunt those regions where their supplies of vegetable food are most abundant. In their train come the predatory kinds which prey upon them. Any change of climate, therefore, unfavourable to the vitality of the pasture will drive away or even locally exterminate the herds of plant-eating animals, and when they disappear the beasts of prey must vanish also.

18. (v.) **Local Influences.**—Various minor causes of a more local kind help to modify the climates of different places, and thereby to affect the flora and fauna. The nature of the soil is one of the most important of these. Wet, marshy ground lowers the mean temperature, seeing that its water absorbs and conveys downward the heat which would otherwise warm the soil. Consequently the effect of drainage is to raise the mean annual temperature. In Britain increase from this cause amounts sometimes to as much as $1\frac{1}{2}°$—$3°$ Fahr., which is as great a change as if the drained ground had actually been transported 100 or 150 miles farther south. A waste of sand presents the greatest extremes of climate, for while the dry surface readily absorbs the sun's heat, so as to rise even to 200° Fahr. during the day, it cools rapidly by radiation, and during a clear night may grow ice-cold.

19. A surface of vegetation prevents the soil from being as much warmed and cooled as it would be if bare, and since leaves never become so hot as soil, they equalise the temperature. A large mass of forest thus exercises a well-marked influence on the climate of the region, tempering alike the heat of the day and the cold of night.

20. Similar effects are produced by lakes. The surface water, chilled by the cold of winter, descends to the bottom, leaving a warmer layer at the top, which, cooled in its turn, sinks down and allows another warmer portion to lie at the surface. By this means the temperature of the air overlying the water is kept above that of the air overlying the adjoining land, while the colder air from all sides flows down to the lake and is there warmed. (Lesson XXVI. Art. 17.) Many deep lakes do not freeze in winter, and then serve as reservoirs of warmth to keep the temperature of the surrounding ground higher than that of places only a short distance away. On the other hand, during summer the water cools the air lying upon it, and thereby lessens the heat of the locality.

21. One other local cause affecting climate may be referred to, viz. proximity to hills or mountains. The influence of high ground shows itself in augmenting rainfall (Lesson X. Art. 29), and in producing currents of air, which, moving alternately up and down the valleys (Lesson XI. Art. 8), give rise to gusts and blasts of cold wind that rush down to the plains.

22. It was formerly imagined that each climate had its own characteristic forms of life, and that the boundaries between the different botanical and zoological regions were as ancient and as well defined as between the various climates. But while similarity of climate does not always bring similarity of vegetation and of animals, the want of resemblance between the plants and animals of two distant countries having similar climates does not arise from any unfitness in the one country for the organisms of the other. Cattle and horses introduced by the Spaniards into South America now run wild there in vast herds. The rat, originally not a native of America, may now be found in all parts of the continent. Hogs, goats, cats, and dogs, first brought into the New World by Columbus and his successors, are to-day found running

wild in great numbers. In Australia, too, the domestic animals introduced by the colonists are rapidly supplanting the kangaroos and other aboriginal forms. A freshwater plant accidentally imported from America has spread rapidly over England, and is choking up canals and the channels of rivers.

23. From the foregoing statements it may be concluded that under similar climates remarkably dissimilar assemblages of plants and animals may exist if they are sufficiently isolated from each other; that such botanical and zoological distribution is not referable to the influence of climate, for plants and animals when artificially removed from the areas within which they are naturally restricted have been found to increase rapidly when transported to a distant but similar climate; and that while climate has evidently an important influence in the distribution of life over the globe, it is not sufficient to account for all that we see.

II.—Migration and Transport.

24. It might be supposed that the present plants and animals first appeared in one region or continent, whence they gradually spread over the whole of the globe. No doubt, many species are endowed with remarkable powers of diffusing themselves, and of living even vigorously under the greatest extremes of climate. But further consideration suffices to convince us that this explanation is wholly incapable of accounting for the existing arrangement of the faunas and floras of the earth.

25. Plants have many facilities for spreading themselves. Their seeds are often swept up into the air by whirlwinds, and may be carried along for hundreds of miles before being dropped again to the ground. Should the soil, climate, and other conditions be favourable, these transported seeds may take root and spread over their

new abode. In other cases, seeds may be borne for long distances over the sea, either floating by themselves or enclosed among earth and leaves in masses of drift-wood. Cast up at last on some remote shore, they sometimes find a fitting home and take root there. To the feathers of birds and the fur of animals seeds must often adhere, and may thus be carried far away from their original source. Seeds which have been for many hours in the crops of birds have been found to be still alive. It may now and then happen, therefore, that when, after flying across hundreds of miles of sea, the birds fall a prey to other predatory members of the feathered race, seeds, falling from their torn crops, find a lodgment in the earth, where eventually they spring into leaf. In these and other ways, many kinds of plants may have spread far beyond their original bounds. Yet, at the best, these are but limited means of transport. Differences of climate and soil, lofty mountain-chains, and intervening seas, have placed barriers in the way of such diffusion, which comparatively few species of plants can ever surmount.

26. Animals enjoy greater facilities for dispersal, since their movements are voluntary as well as involuntary. On some of the large tropical rivers rafts of drift-wood are now and then to be seen, with monkeys and other wild animals upon them, all sailing down the current on their way to the ocean. In the great majority of cases rafts of this kind are broken up at sea, and their unfortunate denizens are drowned. But cases have been known where the animals have actually found their way to land. We may suppose, therefore, that islands in mid-ocean may sometimes have had both plants and animals introduced into them by these means. Again, in the Arctic seas, polar bears have been noticed upon icebergs at a great distance from land; so that by drifting ice as well as by floating vegetation, animals may be diffused from one country to another. But here again the means of transport are so scanty, and the chances of the animals

being able to live and multiply in their new home are so small, that we may be sure it is not in this way that the continents have been peopled.

27. While most animals live within tolerably well-defined limits, marked out by the climate and the kind of vegetation which the climate supports, some species have great powers of migration, and when impelled by their migratory instinct, whether from stress of hunger or from change of season, will travel for hundreds or thousands of miles. In North America, many remarkable instances are on record of the hordes of bisons, beavers, and squirrels, which from time to time have quitted their previous haunts in search of a new home. Birds show this instinct strongly. Many of the most familiar birds of the temperate region, both in the Old and the New World, are migratory. They go north in summer to breed, and after spending some months in a cooler climate, and seeing their young brood able to fly, they again take wing and return to their winter quarters in the south. In Europe the swift, swallow, and cuckoo wing their way in summer even far up within the Arctic Circle, but before winter has set in, they have crossed the Mediterranean to the milder air of Northern Africa.

28. But while a limited number of animals are fitted to spread over wide regions and to endure great diversities of climate, the vast majority are confined within their own district, beyond which they cannot stray, not because they are in all cases unfitted for distant journeys, but because of the insuperable barriers to their advance. Of these obstacles, the most potent is no doubt climate, which acts not only on the animals themselves, but on the vegetation that directly or indirectly furnishes their food. Some species of animals can live only in woods, others cannot stray far from the marshes and jungles, where alone they find their proper support; some are adapted for life solely in the moist hot air and

luxuriant vegetation of the tropics; others find their congenial home among arctic snows.

29. Even, however, where an animal is endowed, like the tiger, with extraordinary powers of accommodating itself to wide extremes of temperature and great variety of food, it is surrounded by many obstacles to its diffusion. A lofty snow-covered mountain-chain may effectually prevent it from crossing into districts where, if it could once reach them, it would find abundance of food and shelter. A strip of barren arid desert is another efficacious barrier, preventing the animals of one province from passing over into another. An arm of the sea or strait only a few miles in breadth suffices to keep the plants and animals of the opposite shores distinct; while of course a wide and deep ocean is an insuperable barrier.

30. Let us suppose, however, that through some exceptionally favourable circumstances, a few animals of one or more species have succeeded in crossing one of these natural barriers, what are the chances that they will be able to establish themselves on the farther side? The climate must be one in which they can live and increase. They must be able to find enough of their proper food. If herbivorous they must needs find vegetation fitted to support them; and if carnivorous they will require to meet with animals less powerful than themselves, in numbers sufficient to yield them subsistence. It would seldom happen that the invaders would have all these chances in their favour. But if they were able to maintain themselves at first, they would soon find their advent opposed by some rival species already long established and numerous. In the struggle that would follow the new comers could seldom make good their hold in the country, and unless able to return to their original haunts would in most cases perish.

31. The vast majority of animals are thus hemmed in by barriers — climate, food, mountains, deserts, wide rivers, seas, or rival species—barriers which, whether seen or unseen, effectually restrain them from spreading beyond the limits of their own district or region. We are therefore forced to conclude that most of the present species of animals (and the same holds true for the great majority of plants) cannot have spread from any one common centre, but from their very nature and requirements must always have been restricted in their distribution, generally to the same tracts in which they now live. While certain forms of plant and animal have been able to diffuse themselves over almost the whole globe, the flora and fauna of each of the great biological regions remain distinctly marked out by more or less definite boundaries. That these regions have in every case had a long history, and that their existing species of plants and animals have been preceded by other and different species, is shown by the rocks which form the land, and the traces both of former vegetable and animal life found in these rocks. In trying to discover how and whence the continents have received their mantle of vegetation and their hosts of animals, science needs to grope backwards among the records contained in the rocks, which form the subject of Geology. To some aspects of this question, which show how closely Physical Geography is linked with Biology, and how the plants and the animals of a continent may be made to tell a part of the ancient history of the land on which they live, the concluding pages of these Lessons may fittingly be given.

III.—Changes of Land and Sea and of Climate.

32. From the facts recorded in previous Lessons it is manifest that the present heights and hollows of the land have not always existed,—that the continents have been uplifted at different times from the bed of the sea,

that each mountain-chain has been ridged up and altered at various periods, and that the valleys have been slowly deepened and widened by the rivers flowing in them. We cannot suppose that such important changes could take place without affecting more or less potently the various forms of plants and animals. Investigation shows that the present distribution of life sometimes bears striking and independent witness to these changes. How this is made evident will be clear from one of the simpler illustrations.

33. The plains of Central Europe up to the shores of the Baltic are clothed with a vegetation which has one common character throughout. Many of the plants are of course local, but a vast number range far and wide over the region. Crossing from the continent into Britain we meet with the same general assemblage of plants, and as the greater number of these could not have been drifted across the intervening sea, but must have travelled by land, they show that originally Britain formed a part of the European continent, and that its separation into islands has taken place since the present species of plants spread over its surface.

34. Any one who ascends the higher hills in Britain, or a part of the mountain-chain of the Alps or the Pyrenees, finds that as he reaches higher elevations he loses the common and characteristic plants of the plains. The vegetation, as it becomes less luxuriant, assumes a more and more distinct type, not only by the disappearance of lowland species, but by the occurrence of others, such as peculiar gentians and saxifrages never seen below, but plentiful on the higher hill-tops and mountain slopes. So general is this change, that every hill or mountain in Central and Northern Europe, rising high enough to reach the fitting climate, may be expected to contain more or less fully this "alpine" flora. And this is found to be the case even when the groups of mountains are separated by wide intervals of low country. The

Scottish Highlands furnish on their higher slopes an abundant growth of alpine forms of vegetation. Crossing the Lowlands, where none of these plants occur, we again meet with them on some of the more elevated summits in the Cheviot Hills. After another interval they reappear on the hill-tops in the Cumberland lake-district, and again on the higher mountains of Wales. Across the whole breadth of England they are absent from the low grounds. They are not to be found on the opposite shores of the Continent. But far to the south they reappear abundantly on the tops of the Pyrenees, and below the snow-line along the whole chain of the Alps.

35. We must not suppose these plants to be merely species peculiar to lofty elevations and always found there. They do not occur on mountains lying to the south of the Palæarctic region. On the Peak of Teneriffe, for instance, they are absent, though the climate and soil would have been well fitted for them. On comparing the heights at which they are met with, we perceive that they approach nearer the sea-level the farther north we trace them. In the Alps they grow in the zone between the upper limit of trees and the snow-line, or at a height of from 6000 to 10,000 feet above the sea. In the Scottish Highlands they descend to 2000 feet, or even lower. In Scandinavia they come down to the sea-level, and grow in such vigour and abundance there as to show that they are really northern or arctic plants.

36. How then did an arctic flora overspread the mountains so far south as the Alps? It could not do so at present. The intervening low grounds are clothed with an abundant vegetation of another type, across which the northern plants could not force their way. To enable them to advance southwards this lowland vegetation would need to be removed. If the climate of Central Europe were to grow as severe as that of Norway or of the higher Alps, the effect of the change would be to kill

the vegetation of the plains, or as much of it as could not endure the greater cold. At the same time the arctic plants, finding their congenial temperature prolonged across the European plain, would gradually spread southwards, descend from the mountains, and finally become the dominant vegetation over the whole region where the arctic climate prevailed. This seems to have been the condition of things when the northern plants found their way to the Alps and Pyrenees.

37. Is there, then, any corroborative evidence that such a wonderful change in the aspect of Europe did really take place? Undoubtedly there is. Below the soil in different parts of the lowlands of France and England, as well as in the deposits covering the floors of caves, bones of reindeer have been found in considerable abundance. This we know is an arctic animal. Remains also of the musk-sheep, glutton, arctic fox, lemming and others, which are all northern forms, have been exhumed from similar situations. So that there can be no doubt that at one time characteristic animals of the arctic regions roamed over Europe as far at least as the south of France. We cannot doubt that this could have happened only through some change of climate which, driving out the usual denizens of the plains and forests, allowed the northern animals to occupy their place.

38. But further and abundant proof of an extremely rigorous climate having overspread Europe is supplied by the polished rocks and heaps of earth described already (Lesson XXVIII.), as part of the work of glacier-ice. Traces of former glaciers are found throughout the more hilly districts of Britain. Similar traces occur both in Norway and among the Alps and Pyrenees, far beyond the limits of the present glaciers, showing that the ice formerly extended in vast sheets into the plains.

39. It was during some part of this cold period that the arctic plants and animals overspread Europe. Since

that time the climate has gradually ameliorated. Step by step, in the struggle for life, the northern plants have been driven out of the plains and up into the mountains, where, among the congenial frosts and snows, to which their competitors on the lower grounds do not follow them, they are able to maintain themselves in scattered colonies. The animals have long since been pushed back into the icy north.

40. In North America a similar record has been preserved. As far south as the White Mountains of New Hampshire in the United States (lat. 45°) the summits are peopled with Labrador plants, which once no doubt extended over the low grounds up into the northern lands, where the same species are now found abundantly down to the sea-level.

41. This illustration—by showing how far the present distribution of plants and animals may be from that which once existed, and also how distinctly groups of plants or of animals may sometimes tell of former changes in physical geography—may serve to indicate why the problem of accounting for the existence and boundaries of the biological regions should be so difficult. So many causes have to be considered and so much knowledge is needed regarding the events which preceded the present state of things. Much has been done by naturalists in this department of research during recent years, but they have as yet only entered upon the beginning of the inquiry. The story of the peopling of each of the great regions with plants and animals may never be fully told. But that it will be made far fuller and clearer than it is, and that it will help to illustrate the history of the continents themselves, cannot be doubted.

42. One grand object of science is to link the present with the past, to show how the condition of the globe to-day is the result of former changes, to trace the progress of the continents back through long ages to their earliest beginnings, to connect the abundant life now teeming in

air, on land, and in the sea, with earlier forms long since extinct, but which all bore their part in the grand onward march of life, now headed by man; and thus, learning ever more and more of that marvellous plan after which this vast world has been framed, to gain a deeper insight into the harmony and beauty of creation, with a yet profounder reverence for Him who made and who upholds it all.

INDEX.

A

ADELSBERG GROTTO, 240
Adriatic Sea, 292
Africa, winds of, 93, 95; coast-line of, 168; average height of, 170; deserts of, 179; plateau character of, 181; coral reefs of, 219; great lakes of, 262, 266; fauna and flora of, 333
Air, composition of, 38; capacity of, for vapour, 43, 65, 80; height of, 45; pressure of, 47; temperature of, 54; moisture of, 64; movements of, 70, 83, 143
Algeria, wells of, 226
Alluvium, 285
Alps, comparative mass of, 170; description of, 174; snow-line of, 177; valleys of, 177; glaciers of, 301, 309, 313; vegetation of, 176, 352
Amazon, River, 290
America, coast-line of, 168; axis of, 171; plains of, 172, 178, 290; salt-lakes of, 173; mountain-chains of, 177; table-lands of, 180; drainage of, 250, 253, 290; water-sheds of, 251; lagoons of, 264; great lakes of, 266
Anchor-ice, 137
Animals, geographical distribution of, 328; migration and transport of, 347
Antarctic Ocean, 109, 114, 129
Anticyclone, 84
Antiparos Grotto, 241
Antipodes, 16
Arabia, desert climate of, 77, 226
Aral, Sea of, 112, 271
Archipelago, 165
Arctic Circle, 15
Arctic Ocean, 109
Arctic regions, ice of, 131, 135; plants and animals of, 330, 332, 336

Artesian wells, 230
Ascension Island, 110
Asia, winds of, 93; position of axis of, 171; salt-lakes of, 173; mountain system of, 178; plateau of, 180
Atlantic Ocean, form and depth of, 109; density of water of, 114; temperature of, 126; height of waves in, 142; currents of, 144; form of, how determined, 165
Atmosphere, 38
Atmospheric pressure, 47, 83, 96
Attraction, force of, 12
Australia, proportion of coast to area of, 168; barrier-reef of, 220; fauna and flora of, 334; animals introduced by man into, 347
Australian region, 334
Avalanches, 299
Azores, 110

B

BALTIC SEA, ground-ice of, 137
Barometer, 48
Bays, 166
Bay of Biscay, sand-dunes of, 101
Beach, origin of a, 152
Beaches, raised, 217
Black Lands of Russia, 179
Black Sea, 113, 295
Blood-rain, 92
Bore of the tidal wave, 150
Borings, internal heat of earth shown by, 17
Brazil-current, 144
Breakers, 142
Britain, rainfall of, 76; winds of, 90, 96; fall of volcanic dust in, 91; sand-dunes of, 101; position of axis of, 171; table-land of, 181; warm springs of, 191, 235

C

CANADA, temperature in, 61; winter in, 137, 295; forests of, 336
Canary Islands, dust showers in, 92
Cancer, Tropic of, 16; Calms of, 90
Cape Horn current, 145
Capricorn, Tropic of, 16; Calms of, 90
Carbon, 41
Carbonic acid, or carbon dioxide, in the air, 41; given off in volcanic districts, 206, 237; solvent power of in water, 237
Caribbean Sea, 110
Carlsbad, hot springs of, 191, 235
Caspian Sea, 112, 173, 179, 270
Catchment-basin, 250
Caverns, formation of by acidulous water, 240; tunnelled, cut by the sea, 158
Chalk, origin of, 186
Challenger expedition, 108-112, 123
Climate, origin of differences of, 60, 156; continental and insular, 156, 342; how determined, 338; affected by distance from Equator, 339; affected by distance from the sea, 341; affected by height above the sea, 342; affected by prevailing winds, 344; affected by local influences, 345; secular changes of, 351
Clouds, formation of, 69; check formation of dew, 68; as indications of aërial currents, 91
Coal, origin of, 186
Coast-lines, 166
Condensation, 66
Conduction, 56
Continents of the globe, 35; distribution of, 164; coast-lines of, 166; general relief of, 169; average height of, 170; axes of, 170; antiquity of, 315
Convection, 56
Coral-reefs, 219
Cotopaxi, 197
Craters, volcanic, 197
Crust of the earth, corrugation of, 209, 315, 317
Currents of the sea, 143
Cyclone, 84

D

DANUBE, annual discharge of, 258; mineral matter transported by, 283, 284

Day, causes in difference of length of the, 13
Dead Sea, 173, 271
Deltas, 247, 291
Deposit by running water, 275
Deserts, climate of, 55, 77, 345; sand wastes of, 101; aspects of, 179
Dew, 68
Dew-point, 68
Diatoms, deposits of, on sea-bottom, 124
Diliquescence, 116
Disease-germs in the air, 39
Distribution of plants and animals, 328
Dolphin Rise, 110
Drainage, effects of, 232, 345
Drainage-basin, 250
Dredge, use of, 108
Dust, importance of, in dry climates, 102

E

EARTH as a planet, 8; proved to be a globe, 8; axis of, 11; movements of, 11; orbit of, 12; hot interior of, 18, 21, 190, 195, 209; history of, 21, 107; measurement and mapping of, 24; diameter of, 29; general view of, 32, 196; size of, 32; density of, 190; possible metallic interior of, 196; contraction of, 107, 209, 315, 317
Earthquakes, 211
Eclipse, 9
Elbe, bore of, 151
Elevation of land. *See* Land
England. *See* Britain
Equator, 11
Equatorial calms, 88; current, 143; regions, plants and animals of, 329, 333, 335
Equinoxes, 13
Erosion by running water, 274, 275
Erratics, 310
Ethiopian region, 333
Europe, winds of, 90, 93, 94, 95, 96; climate of, 155, 342; coast-line of, contrasted with that of Africa, 168; average height of, 170; axis of, 171; plains of, 172, 179; salt-lakes of, 173; table-lands of, 181; watershed of, 250; lagoons of, 264; glaciation of, 313, 354; changes of climate in, 352
Evaporation, 43, 64, 75, 153

INDEX. 359

F

FAROE ISLANDS, 110
Fauna, the assemblage of animals of a district, 331
Field-ice, 133
Firn of Alpine snow-fields, 300
Floe-ice, 133
Flood-plain, 286
Flora, the assemblage of plants of a district, 330; Alpine, of Europe, 352
Florida, peninsula of, 110
Fog, formation of, 69
Föhn, of the Alps, 256
France, dunes of, 101; tidal wave on coast of, 151; hot springs of, 192; extinct volcanoes of, 192, 200; vegetation in, affected by climate, 342
Frost, 79, 294, 323
Fundy Bay, tides of, 150

G

GANGES, drainage area of, 250; mineral matter transported by, 283; delta of, 292, 293
Geneva. *See* Lake of Geneva
Geography, physical, defined, 3
Geological changes influencing climate and the distribution of plants and animals, 351
Geysers, 192
Glaciers, formation of, 131, 299, 300; transport by, 306; erosion by, 311
Globigerina ooze of sea-bottom, 123
Gorges, excavation of, by rivers, 277
Gravity, action of, 12
Great basin of North America, 181
Great Salt Lake of Utah, 269
Greenland, glaciers of, 131, 311; ice-foot of, 135
Grottos, formation of, 240
Ground-ice, 137
Ground-swell, 141
Gulf Stream, 60, 144-154

H

HAIL, 82
"Hard" water, 236
Harmattan, 95
Hawai, 111, 207
Headlands, 166
Heights, measurement of, by barometer, 49
Hemispheres of the Globe, 15, 33; excess of density in Southern, 107; excess of land in Northern, 164
High-water, 148, 152
Himalaya Mountains, height of, 169; height of snow-line on, 80, 343; formation of, 187; glaciers of, 253; vegetation of, 177
Hindostan. *See* India
Hoar-frost, 69
Horizon, extent of the visible, 103

I

ICE, formation of, 78
Icebergs, 129
Ice-foot, 135
Iceland, climate of, 61; volcanic eruptions of, 91; position of, on a submarine plateau, 110; hot springs of, 192
India, rainfall, 76, 99; water-shed of, 251; lagoons of, 264; alluvial plains of, 290
Indian Ocean, temperature of, 127; moisture supplied by, 154
Indus River, 290
Inland seas, 112, 269
Irrawaddy, transport of mineral matter by, 283
Islands, oceanic, of volcanic origin, 111
Isobars, 53
Isotherms, 55
Isthmus, 164

J

JAN MAYEN ISLAND, 207
Japan, winds of, 93; volcanoes of, 207; current, 145
Jupiter, size of the planet, 30
Jura Mountains, 174, 310

K

KARST, honeycombed structure of the, 240
Kentucky, Mammoth Cave of, 241
Khasi Hills, rainfall of, 76, 99
Krakatau, volcanic explosion of, 202, 214

L

LABRADOR, coast ice of, 133; cold currents and climate of, 60, 156
Lagoons, 264
Lake Baikal, 266; Brienz, 288; Como, 267; Geneva, 268, 287; Maggiore, 267; Sabatino, 268; Superior, 266; Thun, 288
Lakes, formation of, 172; without outlet (salt lakes), 172, 181, 262; defined, 258; abundant in Northern latitudes, 259; connected with ice-action, 260, 314; disappearance of, 265, 288; sources of, 265; storms on, 266; depth of, 266; survival of marine forms in, 267; distribution of temperature in, 267; offices of, 268, 287; filter rivers, 287; freezing of, 295, 346; influence of, on climate, 346; revolutions in distribution of, 351
Land, general aspects of the, 162; distribution of, 33, 164; average height of, 170; relief of, 173; nature of materials forming, 182-185; formerly under water, 188; movements of, 210; upheaval of, 212, 216, 316; subsidence of, 213, 218; lowering of level of, by erosion, 284, 320; sculpture of, 314; not original part of the surface of globe, 315
Land-breeze, origin of, 86
Landslips, 242
Latitude, finding the, 27; influence of, on temperature, 58
Lava, nature of, 197; outpouring of, 203; absorbed vapours of, 210
Left bank of a river, meaning of, 240
Life, plant and animal, on the globe, 328
Lime, carbonate of, in sea-water, 118; in springs, 235, 237
Limestone, origin of, 187
Loch Ness, 267; Lomond, 267
Longitude, finding the, 25
Low-water, 148, 152

M

MAELSTROM WHIRLPOOL, 152
Mahanadi River, 226, 254
Malay Archipelago, 207, 333
Map, construction of a, 30
Medicinal springs, 235
Mediterranean Seas, 112

Mercury, distance of planet, from the Sun, 30
Meridians, of longitude, 26; measurement of a degree of, 29
Meteors, 45
Migration of plants and animals, 347
Mines, evidence furnished by, as to internal temperature of the earth, 190
Mississippi, windings of, 249; breadth of, 257; volume of annual discharge of, 257; mineral matter transported by, 283, 284; alluvial plains of, 290; delta of, 291, 292
Missouri, mean slope of, 256
Mist, formation of, 69
Mistral, 94
Moisture, distribution of by winds, 75, 98
Monsoons, 77, 93
Moon, craters of the, 19
Moraines, 307
Mountains, varieties of, 174; formation of, 318; influence of, on climate, 70, 346
Mountain-chains, 173

N

NEAP-TIDES, 148
Nearctic Region, 336
Nebulæ, 23
Nebular theory, 23
Neotropical region, 335
Neptune, planet, distance of, from Sun, 30
Nevé of Alpine snow-fields, 300
Newfoundland, banks of, 110
New Zealand, on a submarine ridge, 111; hot springs of, 194
Niagara, River and Falls, 280
Nile, delta of, 246; annual rise of, 252, 254; alluvium of, 289
Norway, coast-line of, 152, 155; snow-fields and glaciers of, 301, 311
Nova Scotia, 146, 150, 156

O

OASES of deserts, 179, 226
Oceans, 34. *See* Sea
Ooze of sea-bottom, 124
Organic matter in sea-water, 119; in soil, 183
Oriental region, 333
Ozone, 40

P

Pacific Ocean, basin of, 111; depth of, 112; density of water of, 114; deep-sea deposits of, 124; temperature of, 127; tidal wave in, 150; influence of, on climate of America, 154; volcanic girdle of, 206; coral-reefs of, 219
Palæarctic region, 331
Parallels of latitude, 28
Perched blocks, 309
Persian Gulf, filling up of, 290
Peru, rainless climate of, 77; elevation of, 217
Plains of the globe, 178; formed by alluvial deposits, 326
Planets, 22
Plants, distribution of, according to height, 176; geographical distribution of, 328; migration and transport of, 347; remains of in rocks, 186
Plateaux, 180
Po, delta of, 291
Polar regions, plants and animals of, 330, 332, 337
Pole the Celestial, 27
Poles, North and South, 11
Pompeii, buried under the ashes of Vesuvius, 199, 202
Pot-holes, 277
Prairies, 178
Precipitation, zone of Constant, 76, 88
Protoplasm, in sea-water, 119
Pumice of sea-bottom, 123
Pyrenees, arctic vegetation of, 352

R

Race of the tides, 151
Radiation, 56; checked by water-vapour in the air, 67
Radiolaria, deposits of, 124
Rain, washes the air, 40, 78; formation of, 74; composition of, 78, 114, 233; floats on sea-water, 115; supplies springs, 224; work done by, 272
Rainfall, distribution of, 76, 153, 282; proportion of, carried to sea by rivers, 253; of Britain, 76; Caspian basin, 113; Europe, 77; India, 76, 77; South America, 77
Rainless climates, 77, 154, 179, 255
Rain-prints, 273
Rainy seasons, 16, 254
Raised beaches, 217
Rapids, 279
Revolution of the earth, 12
Rhine, drainage area of, 250; annual rise of, 255; mineral matter in solution in water of, 281
Rhone, mineral matter transported in water of, 281, 284
Right bank of a river, meaning of, 249
Rivers, underground, 232, 240; course of typical, 245; sources of, 252; flow, 256; volume of water discharged by, 257; erosion by, 275; transporting power of, 280; deposits from, 285; freezing of, 295
River-bars, 290
River-basins, 244
Rocks, bedded, 185, 316; containing plant remains, 186; containing animal remains, 186; crystalline, 188; specific gravity of, 190
Rocky Mountains, 177, 178, 187, 337
Rotation of the earth, 11

S

Sahara, Desert of, 77, 101, 179, 218, 226, 333
St. Helena, height of tide at, 150
St. Paul's Rock, Atlantic Ocean, 110
Salt, incrustations of, 263
Salt-lakes 262, 269
Salts, of sea-water, 115
Sand-dunes, 100
Sandstone, origin of, 185
Sandwich Islands, 111, 203, 207
Sargasso Sea, 145
Scandinavia, oscillation of level in, 218; position of axis of, 171; table-land of, 181
Scenery, how influenced by sculpture of the land, 320
Sea, general aspects of the, 104; depth of, 106, 110, 111, 112; deepest abysses of, 112; saltness of, 113; density of water of, 114; composition of water of, 117-119; nature of floor of, 120, 165; plants and animals on bottom of, 120; temperature of, 125; ice of, 129; movements, 137; offices of, 152; supplies the atmosphere with moisture, 153; regulates the distribution of temperature, 60, 154; wears away its shores, and thus

tends to reduce the area of the dry land, 156; receives and preserves the materials out of which new land will be formed, 160; action of, on the whole conservative, 161; submarine ridges and oceanic islands of, 165
Sea-basins of the globe, 103
Sea-breeze, origin of, 86
Sea-dust, 92
Sea-level, 35
Sea-shore, forms of, 166; waste of, 156; gain of land at, 160
Seasons, cause of the, 15
Sebka-el-Faroon, 256
Seine, bore of the, 151
Shannon, drainage area of, 250
Siberian tundras, 179, 332
Silica in sea-water, 117
Silvas of South America, 290
Simoom, 95
Sinter of hot springs, 192
Sirocco, 94
Sleet, 82
Snow, 78, 298
Snow-fields, 298
Snow-line, 81, 177, 343
Sodium-chloride in sea-water, 116
"Soft" water, 236
Soil, formation of, 182; influence of, on climate, 345
Solano, 95
Solar System, 23
Sounding-line, use of, 108
Space, vastness of, 1
Spain, table-land of, 181; hot springs of, 191
Spitzbergen, 144
Spring-tides, 148
Springs, 222; dependent on rainfall, 224; surface, 227; deep-seated, 228; composition of water of, 233; mineral, 235; temperature of, 235; thermal, 191, 236; quantity of mineral matter abstracted by, 237; origin of substances dissolved by, 237
Stalactites, origin of, 238
Steppes, 179
Storms, 95; how caused, 53; destructiveness of, 100
Stratification of rocks, 185
Striation of rocks by ice, 312
Subsidence of land, 173, 218
Subsoil, formation of, 184
Sun, distance of, 10, 12, 30; heat of, 21, 56, 64; composition of the, 22; rotation of, 22; size of, 30
Sun-spots and weather, 63

T

TABLE-LANDS, 180
Tahiti, 111
Tay, drainage area of, 250; annual discharge of, 258
Temperate regions, plants and animals of, 329, 332, 337, 340
Temperature, how determined, 54; how interchanged, 56; regulated by latitude, 58; regulated by form of land and height above sea, 63; daily range of, 63; distribution of by winds, 97
Teneriffe, Peak of, 198, 353
Thames, drainage area of, 250; mineral matter in water of, 281; frozen over, 295
Thermometer, 54, 108
Tiber, delta of, 291
Tides, 139, 147
Tigris River, 290
Trade-winds, 90
Transport by running water, 275, 280; of plants and animals, 347
Travelled blocks, 310
Triangulation, 31
Tristan d'Acunha, 110
Tropical regions, rainfall of, 76, 153; fauna and flora of, 177, 329, 333, 336
Tropics, 16
Tundras, 179, 270
Tuscarora expedition, 111, 112

U

UNDERGROUND circulation of water, 222
United States, winds of, 89, 94; mountains of, 177, 180; plains of, 178; salt-lakes of, 173; table-lands of, 180; geysers of, 194; volcanic action in, 206; earthquakes in, 213; fauna and flora of, 336, 341, 349, 355
Upheaval of land, 317
Utah, Great Salt Lake of, 269

V

VALLEYS, longitudinal and transverse, 177; hollowed out by running water, 325; deepened by glaciers, 312

INDEX. 363

Vegetation, influence of, upon climate, 345; distribution of, regulated by climate, 338
Vesuvius, eruptions of, 108, 201, 203, 204, 205
Volcanic dust, transported by aërial currents, 91; eruption, 198
Volcanoes, nature of, 194, 196
Volga, mean slope of, 256, 270

W

WADYS OF SYRIA, 255
Water, circulation of, on the globe, 44, 114, 222; its three conditions, 68; underground circulation of, 224; possible ultimate abstraction of from the surface, 227; hard and soft, 234; work of running, 272, 325; point of maximum density of, 294
Water-courses, 244
Waterfalls, 277
Water-shed, 250
Water-vapour in the air, 43, 56, 64, 86
Water-worn character of detritus, 275

Waves, formed by wind, 140; height of, 142; force of, 142, 157; caused by earthquakes, 213
Weather forecasts, 95
Wells, 225, 230
West Indian Islands, 110; plants drifted from, to shores of Europe, 139
Westward growth of European towns, cause of, 90
Whirlpools, 152
Wiesbaden, hot springs of, 191, 235
Winds, cause of, 85; periodical (seasonal), 77, 93; constant, 88; local, 94; rate of, in storms, 96; office of, 97, 118, 256; wet and dry, 99

Y

YEAR, how determined, 12
Yellowstone, geysers of the, 194

Z

ZENITH, 28
Zirknitz, Lake of, 241, 265

THE END.

Printed by R & R. CLARK, *Edinburgh.*

MESSRS. MACMILLAN AND CO.'S PUBLICATIONS.

BY THE SAME AUTHOR.

TEXT-BOOK OF GEOLOGY. By Sir ARCHIBALD GEIKIE, LL.D., F.R.S.; Director-General of the Geological Survey of Great Britain and Ireland, and Director of the Museum of Practical Geology London; lately Murchison Professor of Geology and Mineralogy in the University of Edinburgh, and Director of the Geological Survey of Scotland. With Illustrations. Second Edition. Med. 8vo. 28s.

TIMES—"The Text-book in our estimation contains a more complete account of the science of geology, in its latest developments, than is to be found in any similar English work.... In a clear, straightforward, and methodical manner he brings out the principles of his science, traces them through all their ramifications, and marshals all his facts and deductions. This is as it should be."

THE SCENERY OF SCOTLAND VIEWED IN CONNEXION WITH ITS PHYSICAL GEOLOGY. With a geological map and illustrations. By the same Author. Second Edition. Crown 8vo. 12s. 6d.

CLASS-BOOK OF GEOLOGY. By the same Author. Illustrated with woodcuts. Second Edition. Crown 8vo. 4s. 6d.

ELEMENTARY LESSONS IN PHYSICAL GEOGRAPHY. Illustrated with woodcuts and ten plates. By the same Author. Fcap. 8vo. 4s. 6d.

*** QUESTIONS. *For the use of Schools.* Fcap. 8vo. 1s. 6d.

PHYSICAL GEOGRAPHY. By the same Author. With Illustrations. Pott 8vo. 1s. [*Science Primers.*

GEOLOGY. By the same Author. With Illustrations. Pott 8vo. 1s. [*Science Primers.*

OUTLINES OF FIELD GEOLOGY. By the same Author. New and revised Edition. Extra Fcap. 8vo. 3s. 6d.

GEOLOGICAL SKETCHES AT HOME AND ABROAD. By the same Author. With Illustrations. 8vo. 10s. 6d.

MACMILLAN AND CO., LONDON.

MACMILLAN'S GEOGRAPHICAL SERIES.

EDITED BY SIR ARCHIBALD GEIKIE, F.R.S.,
Director-General of the Geological Survey of the United Kingdom.

THE TEACHING OF GEOGRAPHY. A Practical Handbook for the use of Teachers. By Sir A. GEIKIE, F.R.S. Globe 8vo. 2s.

GEOGRAPHY OF THE BRITISH ISLES. By Sir ARCHIBALD GEIKIE, F.R.S. Pott 8vo. 1s.

THE ELEMENTARY SCHOOL ATLAS. Twenty-four maps in colours. By JOHN BARTHOLOMEW, F.R.G.S. 4to. 1s.

AN ELEMENTARY CLASS-BOOK OF GENERAL GEOGRAPHY. By HUGH ROBERT MILL, D.Sc. Edin. Illustrated. Crown 8vo. 3s. 6d.

MAPS AND MAP DRAWING. By W. A. ELDERTON, Pott 8vo. 1s.

GEOGRAPHY OF EUROPE. By JAMES SIME, M.A. With Illustrations. Globe 8vo. 3s.

ATHENÆUM—"It is quite equal in point of execution to Prof. A. Geikie's *Geography of the British Isles*. Due prominence has been given to the physical features of each country, and the author has very wisely not eschewed historical references, so far as they are related to geography."

GLASGOW HERALD—"It contains a vast amount of remarkably well-arranged information, and numerous illustrations are given both of the more striking European cities and of natural phenomena."

DUBLIN EVENING MAIL—"For a general geography of Europe, this is the most interesting we have ever read."

SCOTTISH LEADER—"It illustrates the great progress made during the last few years in the methods of instruction in this department of knowledge. Mr. Sime furnishes a picturesque description of the physical aspects and geological history, first of the whole continent of Europe and then of each separate country. In the same way he outlines the race-history, so to speak, of each people, and then epitomises the story of their political, social, industrial, and commercial development. A series of excellent woodcuts add to the point and interest of the text, and help to make the book a manual such as any child of average intelligence and healthy tastes can study with positive pleasure."

ELEMENTARY GEOGRAPHY OF INDIA, BURMA, AND CEYLON. By H. F. BLANFORD, F.G.S. Globe 8vo. 2s. 6d.

ELEMENTARY GEOGRAPHY OF THE BRITISH COLONIES. By G. M. DAWSON and A. SUTHERLAND. Globe 8vo. 3s.

GEOGRAPHY OF NORTH AMERICA. By Prof. N. S. SHALER. [*In the Press.*

MACMILLAN AND CO., LONDON.

MESSRS. MACMILLAN & CO.'S GEOGRAPHICAL BOOKS.

THE ELEMENTARY SCHOOL ATLAS. By JOHN BARTHOLOMEW, F.R.G.S. 4to. 1s.

MACMILLAN'S SCHOOL ATLAS, PHYSICAL AND POLITICAL. 80 Maps and Index. By the same. Royal 4to. 8s. 6d. Half-morocco, 10s. 6d.

PROCEEDINGS OF THE ROYAL GEOGRAPHICAL SOCIETY says—"The maps are all very nicely drawn, and well suited to the purpose for which they have been published. Among others the large map of the world on Mercator's projection is worthy of special commendation, as are also the maps of Africa, which have been carefully brought up to date. In addition to a diagram illustrating the vertical distribution of climate, an excellent sheet of the projections most frequently used in the construction of maps, and one on which the solar system, the seasons, eclipses, etc., are shown, there are sixty sheets of physical and political maps."

SCOTTISH GEOGRAPHICAL MAGAZINE says—"This Atlas should meet all the requirements of schools. The selection of maps is a happy one, due regard having been given to British interests. The maps are clearly printed, and are not overcrowded with names. It is satisfactory to notice that, in the general index, latitudes and longitudes are given in every case. We can confidently recommend this Atlas for use in schools."

THE LIBRARY REFERENCE ATLAS OF THE WORLD. By the same. 84 Maps and Index to 100,000 places. Half-morocco. Gilt edges. Folio. £2 12s. 6d. net. Also in parts, 5s. each, net. Index, 7s. 6d. net.

CLASS-BOOK OF GEOGRAPHY. By C. B. CLARKE, F.R.S. With 18 Maps. Fcap. 8vo. 3s.; sewed, 2s. 6d.

A SHORT GEOGRAPHY OF THE BRITISH ISLANDS. By JOHN RICHARD GREEN, LL.D., and A. S. GREEN. With Maps. Fcap. 8vo. 3s. 6d.

A PRIMER OF GEOGRAPHY. By Sir GEORGE GROVE. 18mo. 1s.

A MANUAL OF ANCIENT GEOGRAPHY. By Dr. H. KIEPERT. Crown 8vo. 5s.

LECTURES ON GEOGRAPHY. By General RICHARD STRACHEY, R.E. Crown 8vo. 4s. 6d.

A PRIMER OF CLASSICAL GEOGRAPHY. By H. F. TOZER, M.A. 18mo. 1s.

MACMILLAN AND CO., LONDON.

MESSRS. MACMILLAN AND CO.'S PUBLICATIONS.

THE RUDIMENTS OF PHYSICAL GEOGRAPHY FOR INDIAN SCHOOLS; with Glossary. By H. F. BLANFORD, F.G.S. Crown 8vo. 2s. 6d.

A POPULAR TREATISE ON THE WINDS. Comprising the General Motions of the Atmosphere, Monsoons, Cyclones, etc. By W. FERREL, M.A., Member of the American National Academy of Sciences. 8vo. 18s.

PHYSICS OF THE EARTH'S CRUST. By Rev. OSMOND FISHER, M.A., F.G.S., Hon. Fellow of King's College, London. Second Edition, enlarged. 8vo. 12s.

PHYSIOGRAPHY. An Introduction to the Study of Nature. By T. H. HUXLEY, F.R.S. Illustrated. Crown 8vo. 6s.

SPECTATOR—"The book will be invaluable in producing in young people an interest in the phenomena of nature. It is not a 'hard' book; the subjects are treated simply and, it is needless to add, accurately, and all technical terms are explained when they are first used, the words from which they are derived being given in footnotes. The work will also be useful to teachers as a model of the method of instruction."

ACADEMY—"It would hardly be possible to place a more useful or suggestive book in the hands of learners and teachers, or one that is better calculated to make Physiography a favourite subject in the Science Schools."

SATURDAY REVIEW—"They are written in that attractive style which is characteristic of a great natural history demonstrator, a style in which clearness and precision of language are combined with a vivid survey of the various objects touched upon."

OUTLINES OF PHYSIOGRAPHY—THE MOVEMENTS OF THE EARTH. By J. NORMAN LOCKYER, F.R.S., Examiner in Physiography for the Science and Art Department. Illustrated. Cr. 8vo. Sewed, 1s. 6d.

MACMILLAN AND CO., LONDON.

MACMILLAN'S SCIENCE CLASS-BOOKS.

Fcap. 8vo.

LESSONS IN APPLIED MECHANICS. By J. H. COTTERILL and J. H. SLADE. 5s. 6d.

LESSONS IN ELEMENTARY PHYSICS. By Prof. BALFOUR STEWART, F.R.S. New Edition. 4s. 6d. (Questions on, 2s.)

EXAMPLES IN PHYSICS. By Prof. D. E. JONES, B.Sc. 3s. 6d.

ELEMENTARY LESSONS ON SOUND. By Dr. W. H. STONE. 3s. 6d.

AN ELEMENTARY TREATISE ON STEAM. By Prof. J. PERRY, C.E. 4s. 6d.

ELEMENTARY LESSONS IN ELECTRICITY AND MAGNETISM. By Prof. SILVANUS P. THOMPSON. 4s. 6d.

ABSOLUTE MEASUREMENTS IN ELECTRICITY AND MAGNETISM. By A. GRAY, F.R.S.E. 5s. 6d.

POPULAR ASTRONOMY. By Sir G. B. AIRY, K.C.B., late Astronomer-Royal. 4s. 6d.

ELEMENTARY LESSONS ON ASTRONOMY. By J. N. LOCKYER, F.R.S. New Edition. 5s. 6d. (Questions on, 1s. 6d.)

LESSONS IN ELEMENTARY CHEMISTRY. By Sir H. ROSCOE, F.R.S. 4s. 6d.

PROBLEMS ADAPTED TO THE ABOVE. By Prof. THORPE and W. TATE. With Key. 2s.

OWENS COLLEGE JUNIOR COURSE OF PRACTICAL CHEMISTRY. By F. JONES. With Preface by Sir H. ROSCOE, F.R.S. 2s. 6d.

MACMILLAN'S SCIENCE CLASS-BOOKS—Continued.

OWENS COLLEGE COURSE OF PRACTICAL ORGANIC CHEMISTRY. By Julius B. Cohen, Ph.D. With Preface by Sir H. Roscoe and Prof. Schorlemmer. 2s. 6d.

ELEMENTS OF CHEMISTRY. By Prof. Ira Remsen. 2s. 6d.

CHEMICAL THEORY FOR BEGINNERS. By L. Dobbin, Ph.D., and J. Walker, Ph.D. 2s. 6d.

LESSONS IN ELEMENTARY PHYSIOLOGY. By Rt. Hon. T. H. Huxley, F.R.S. 4s. 6d. (Questions on, 1s. 6d.)

LESSONS IN ELEMENTARY ANATOMY. By St. G. Mivart, F.R.S. 6s. 6d.

LESSONS IN ELEMENTARY BOTANY. By Prof. D. Oliver, F.R.S. 4s. 6d.

ELEMENTARY LESSONS IN THE SCIENCE OF AGRICULTURAL PRACTICE. By Prof. H. Tanner. 3s. 6d.

DISEASES OF FIELD AND GARDEN CROPS. By W. G. Smith. 4s. 6d.

LESSONS IN LOGIC, INDUCTIVE AND DEDUCTIVE. By W. S. Jevons, LL.D. 3s. 6d.

POLITICAL ECONOMY FOR BEGINNERS. By Mrs. Fawcett. With Questions. 2s. 6d.

ELEMENTARY LESSONS IN PHYSICAL GEOGRAPHY. By Sir Archibald Geikie, F.R.S. 4s. 6d. (Questions on, 1s. 6d.)

CLASS-BOOK OF GEOGRAPHY. By C. B. Clarke, F.R.S. 3s.; sewed, 2s. 6d.

HANDBOOK OF PUBLIC HEALTH AND DEMOGRAPHY. By Ed. F. Willoughby, M.B. 4s. 6d.

MACMILLAN AND CO. 15.9.93.

www.ingramcontent.com/pod-product-compliance
Lightning Source LLC
Chambersburg PA
CBHW020100020526
44112CB00032B/585